T0331886

Modeling Gravity Hazards from Rockfalls to Landslides

Discrete Granular Mechanics Set

coordinated by
Félix Darve

Modeling Gravity Hazards from Rockfalls to Landslides

Vincent Richefeu
Pascal Villard

First published 2016 in Great Britain and the United States by ISTE Press Ltd and Elsevier Ltd

ISTE Press Ltd
27-37 St George's Road
London SW19 4EU
UK

www.iste.co.uk

Elsevier Ltd
The Boulevard, Langford Lane
Kidlington, Oxford, OX5 1GB
UK

www.elsevier.com

British Library Cataloguing-in-Publication Data
A CIP record for this book is available from the British Library
Library of Congress Cataloging in Publication Data
A catalog record for this book is available from the Library of Congress
ISBN 978-1-78548-076-8

Printed and bound in the UK and US

Contents

Foreword

Molecular dynamics is recognized as a powerful method in modern computational physics. This method is essentially based on a factual observation: the apparent strong complexity and extreme variety of natural phenomena are not due to the intrinsic complexity of the physical laws but due to the very large number of basic elements in interaction through, in fact, simple laws. This is particularly true for granular materials in which a single intergranular friction coefficient between rigid grains is enough to simulate, at a macroscopic scale, the very intricate behavior of sand with a Mohr–Coulomb plasticity criterion, a dilatant behavior under shearing, non-associate plastic strains, etc. and, *in fine*, an incrementally nonlinear constitutive relation. Passing in a natural way from the grain scale to the sample scale, the discrete element method (DEM) is able to precisely bridge the gap between micro and macro-scales in a very realistic way, as it is today verified in many mechanics labs.

Thus, DEM is today in an impetuous development in geomechanics and in the other scientific and technical fields related to grain manipulation. Here lies the basic reason for this new set of books called "Discrete Granular Mechanics", in which not only are numerical questions considered but also experimental, theoretical and analytical aspects in relation to the discrete nature of granular media. Indeed, from an experimental point of view, computational tomography, for example, DEM is giving rise today to the description of all the translations and rotations of a few thousand grains inside a given sample and to the identification of the formation of mesostructures such as force chains and force loops. With respect to theoretical aspects, DEM is also confirming, informing or at least precising some theoretical clues such as the questions of failure modes, of the expression of stresses inside a partially saturated medium and of the mechanisms involved in granular avalanches. Effectively, this set has been planned to cover all the experimental, theoretical and numerical approaches related to discrete granular mechanics.

The observations show undoubtedly that granular materials have a double nature, that is continuous and discrete. Indeed, roughly speaking, these media respect the matter continuity at a macroscopic scale, whereas they are essentially discrete at the granular microscopic scale. However, it appears that, even at the macroscopic scale, the discrete aspect is still present. An emblematic example is constituted by the question of shear band thickness. In the framework of continuum mechanics, it is well recognized that this thickness can be obtained only by introducing a so-called "internal length" through "enriched" continua. However, this internal length does not seem to be intrinsic and to constitute a kind a constitutive relation by itself. Probably, it is because considering the discrete nature of the medium by a simple scalar is oversimplifying reality. However, in DEM modeling, this thickness is obtained in a natural way without any *ad hoc* assumption. Another point, whose proper description was indomitable in a continuum mechanics approach, is the post-failure behavior. The finite element method, which is essentially based on the inversion of a stiffness matrix, which is becoming singular at a failure state, meets some numerical difficulties when going beyond a failure state. Here, also, it appears that DEM is able to simulate fragile, ductile, localized or diffuse failure modes in a direct and realistic way – even in some extreme cases such as fragmentation rupture.

The main limitation of DEM is probably linked to the limited number of grains or particles which can be considered while maintaining an acceptable computation time. Thus, the simulation of boundary value problems remains bounded by more or less heuristic cases. So, the current computations in labs involve at best a few hundred thousand grains and, for specific problems, a few million. Let us note, however, that the parallelization of DEM codes has given rise to some computations involving 10 billion grains, significantly broadening the field of applications for the future.

In addition, this set of books will also present recent developments in micromechanics, applied to granular assemblies. The classical schemes consider a representative element volume. These schemes propose to go from the macro-strain to the displacement field by a localization operator, then the local intergranular law relates the incremental force field to this incremental displacement field, and eventually a homogenization operator deduces the macro-stress tensor from this force field. The other possibility is to pass from the macro-stress to the macro-strain by considering a reverse path. So, some macroscopic constitutive relations can be established, which properly consider an intergranular incremental law. The greatest advantage of these micromechanical relations is probably to consider only a few material parameters, each one with a clear physical meaning. This set of around 20 books has been envisaged as an overview of all the promising future developments mentioned earlier.

Félix DARVE
July 2016

Introduction

A number of infrastructures are built in mountainous areas where gravity hazards are of major concern. For people who live and work there, the general feeling of safety is certainly based on the fact that most hazards seem satisfactorily handled by the engineers of the field. Besides, a certain anticipation is perceptible when people see the great number of protective structures that are easily identifiable in mountainous regions. The valuable expertise of the engineers is, roughly speaking, concentrated around three questions:

1) May a gravitational hazard occur?

2) What would be the consequences in terms of impacted areas?

3) How can infrastructures – and human life – be protected from these events?

To address these questions, engineers support their expertise with (numerical) tools that must be easy to use and able to provide rapid and helpful hints. Most often, these tools are trajectory analysis for fragmental rockfalls, and they are shallow-layer models for large-scale rock avalanches or landslides. For medium-sized rock avalanches, no specific dedicated tool is used; most of time the trajectory analysis is still employed with no awareness of the dissipation that occurs in between the rock blocks themselves.

It is the great expertise of engineers that achieves the goal of addressing the above three questions – because the problem dealt with is extremely complex. As researchers, however, it is possible to put aside the context of "risk management" to focus on more scientific aspects. Consequently, the present work outlines some numerical tools intended more for academic research than for routine use in an engineering framework. Note, however, that these tools have been developed to deal with issues requested by stakeholders. The use of such tools might be ordinary in the future, the purpose being to make them reliable and robust so that a beginner can operate them.

It should be noted that this book will only focus on the propagation (flows and stops) of masses without taking an interest of the initiation of these mass flows – that is the cause of its destabilization. In other words, the first question above is not addressed at all, nor the third (even if some protective structures will be considered a couple of times in the simulations presented). Basically, all simulations will consist of the release of an already unstable mass that will flow over more or less sophisticated terrain due to gravity. The termination of the propagation will result from *how* and *where* the dissipation occurred.

In the framework of the *discrete element method* (DEM) that will be widely reported in this book, all blocks interact while their trajectories obey Newton's laws. The dissipation is thus introduced at the collision or the persisting contact level. In the field of trajectory analysis, which is widely used when dealing with rockfalls, the dissipation is controlled by the so-called coefficient of restitution (CoR) that expresses how much energy (or velocity) remains after a collision.

Most of time, these CoR encompass loss of energy from different origins; for instance, the CoR in the tangential direction of the collision plane may include the loss of energy due to friction together with the loss or gain of kinetic energy caused by the angular velocity of the block before the collision. From the very beginning of the work, a particular choice, which is inherent to DEM simulations, has been followed: the kinematic effects have been separated from the dissipation at the contacts. The dissipation only occurs by means of the balance of the force works, and the kinetics changes as a result. In other word, the dissipation concerns the contacts/collisions and not entire blocks; this is different from the trajectory analysis that usually treats the block as a point where the mass has been lumped. Without saying anything about the physics concretely implied at the collision (e.g. local damage or breakage, elastic wave absorption, viscosity), two inputs are distinguished in the dissipated energy for a given collision: the dissipation due to the forces acting in the direction normal to the contact and the dissipation of the tangential forces. A third type of dissipation involves a resistance moment at the contact point when a block collides with a soft substratum. All these dissipation terms are tracked by means of the works of the contact forces (or moment) by assuming that the works of the elastic components are negligible (this is indeed a weak assumption). When considering the evolution of the forces acting at each contact in a DEM simulation, a complete mapping of the loss of energy can be drawn both in space and during time. The DEM associated with this analysis framework provides a powerful tool with a rich database for analysis.

In addition to the dissipation analysis, the core of the approach relies on the shape of the blocks and, for a second time, on the topology of the propagation terrain. It is believed that these geometric features are key elements that play a crucial role in the global behavior of a granular flow. We will see, throughout the reading, that this is indeed the case. However, no attempt to characterize the shape will be proposed. Instead, the analysis will be built upon more evocative descriptions such as "the propensity of a block for rolling or for sliding" or the "disturbance of a granular flow due to the bumpyness of the terrain".

Another notable aspect of the work is the fact that all simulations are deterministic. Indeed, we believe that the explicit consideration of geometries (for both the blocks and the terrain) eliminates the need for using some stochastic inputs typically used in rockfall simulations. For example, the explicit use of a complex shape will naturally involve a statistical distribution of the deviation angles after a block collision on a slope; the rate of energy dissipation will also obey a certain distribution as a function of the configuration of collision (in particular, the angular velocity and position of the block relative to the slope) with no need to explicitly define this distribution. Once again, the role of geometry is shown to be decisive and the bias is to say that "an accurate geometry is better than lots of parameters".

Despite all recognition enjoyed by DEM, this approach mainly faces two difficulties: (1) the mechanical parameters can be difficult to assess and (2) the computation duration becomes prohibitive when the number of elements is too large. This latter issue is canceled when assuming the granular medium as a continuum. Although many benefits are lost, the model becomes usable in the framework of engineering. For that reason, the book also deals with the *material point method* (MPM), which is a relatively new and promising approach.

It is important, when reading this book, to foster a mindset of truth seeking rather than targeting the applicability of the different models depicted. Certainly, this attitude of curiosity is helpful for a better appreciation of the messages delivered – the authors are evidently well aware about the simplifications made.

This book is composed of five chapters. Chapter 1 provides a description of the employed models. The validity of the specific DEM developed will then be dealt with – in Chapter 2 – by means of comparisons with release experiments. The framework of typical analysis will also be presented. Chapter 3 discusses the influences of several mechanical and geometrical parameters that can be involved in the propagation processes. Chapter 4 will present three concrete cases of medium rock avalanches that implement sophisticated *digital surface models* together with block shapes issued from natural discontinuity plans (or blasting). Finally, a comparison of mass releases simulated by DEM and MPM will be introduced in Chapter 5.

The authors would like to insist upon one point very strongly: the studies presented in the book result from the work of several key players – so-called co-workers – who are colleagues or students. The names of these co-workers have been mentioned at the beginning of the chapters where their work has been involved.

1

Computational Methods

In this chapter, computational methods will be described with a particular focus on the discrete element method (DEM). The question of knowing which method is most appropriate for one or another situation is not directly addressed here. Instead, the fundamental methods are discussed from both the physical and the computation point of view.

1.1. Trajectory analysis

The aim in this section is to provide the basic concepts of the approach without going too deep into considerations. In reality, there are so many variants that one cannot focus on a particular one.

Basically, the main strategy to conduct a trajectory analysis is the *lumped mass* approach that treats a block as a point with a mass [RIT 63, PIT 76, HUN 88]. More recent approaches deal with shaped blocks [FAL 85, DES 87]; they are not reported here for the sake of brevity. A trajectory analysis consists of generating a high number of trajectories by varying the initial conditions and by introducing some controlled randomness in the collisions with the terrain. Nowadays, this randomness is almost always accounted for in the trajectory models [CRO 04, DOR 04, BOU 09a, CHR 07] used by the stakeholders. In this way, it is possible to make a prediction on where the blocks shall stop (zoning), their more probable paths, their maximum height at a given position and so on.

The trajectories themselves are constructed by alternating free flight phases and collisions with the terrain. The friction of air is usually neglected during the free flights so that the path – given by $\mathbf{v}(t)$ and $\mathbf{x}(t)$ – of a material point (MP) starting at position $\mathbf{x}_0$ with velocity $\mathbf{v}_0$ is:

$$\mathbf{v}(t) = g\mathbf{z}t + \mathbf{v}_0 \quad \text{and} \quad \mathbf{x}(t) = \frac{g\mathbf{z}}{2}t^2 + \mathbf{v}_0 t + \mathbf{x}_0 \qquad [1.1]$$

where $g = -9.81$ ms^{-2} is the gravity acceleration and $\mathbf{z}$ is a unit upward vector.

Each collision dissipates energy as a function of the impact angle, the slop and the velocity. Considering the collision velocities along the direction normal and tangential to the slope, v_n and v_t, respectively, two parameters are generally used to distinguish the dissipation in these directions in terms of the ratio of velocities after and before the collision – superscripted ($^+$) and ($^-$), respectively. The most common expression of these coefficients of restitution reads:

$$e_n = -\frac{v_n^+}{v_n^-} \quad \text{and} \quad e_t = \frac{v_t^+}{v_t^-} \qquad [1.2]$$

When an intersection of a trajectory with the terrain is detected (for increasing time), the normal and tangential velocities v_n^- and v_t^- before the collision are used to compute the velocities after the collision:

$$v_n^+ = -e_n v_n^- \quad \text{and} \quad v_t^+ = e_t v_t^- \qquad [1.3]$$

The collision position together with these "reflected" velocities are used as initial conditions for the next free flight phase. Doing so, the collision duration is zero – even if the collision duration can be accounted for in the collision model.

Besides, the definition of the coefficient of restitution is the core of the model, and as a matter of fact, many different definitions are used. They are more sophisticated and may utilize geometric, energetic or stochastic considerations. Also, it can be considered as a coupled effect of the normal and tangential velocities so that the velocities after collision are expressed that way:

$$\begin{pmatrix} v_n^+ \\ v_t^+ \end{pmatrix} = \begin{bmatrix} e_{nn} & e_{nt} \\ e_{nt} & e_{tt} \end{bmatrix} \cdot \begin{pmatrix} v_n^- \\ v_t^- \end{pmatrix} \quad [1.4]$$

In tree-dimensional simulations, the tangential velocities is a vector $\mathbf{v}_t$ and some stochasticity can be introduced, in particular in the angle between the vertical planes holding the trajectories before and after the instant of collision [BOU 09b].

In the end, the trajectory analysis models focus on the collision behavior. Each model has its own recipe, which is more or less sophisticated, but the collisions occur only between MPs (or sometimes rigid shaped blocks) and the terrain. No interaction is taken into consideration in-between the MPs themselves.

1.2. Discrete element method

The "classical" DEM is sketched in this section as a specific application to rock avalanches. This is the DEM with explicit time integration and force laws expressed as an explicit relation between contact force and local variables. Although many references exist on the topic of smooth DEM, most of them focus on quasistatic loadings and/or spherical bodies. This section presents a numerical model dedicated to the flow of rigid blocks on a rigid terrain, implemented in a C++ toolkit called `DEMbox`.

1.2.1. *Block shapes*

For rockfall modeling, the geometry of the blocks has been chosen so that it conforms the actual block shapes. The approach of *sphero-polyhedra* (SP) has thus been adopted. In essence, the principle is similar to that of "clumps" where spheres are assembled to form a rigid object of complex shape. With SP, in addition to spheres, two extra forms are assembled: cylinders to form the edges and planar polygons to form the faces. The vertexes are formed by spheres. Figure 1.1 provides an overview of an SP representing a pebble, which has the particularity of being non-convex. The body shape is actually defined by sweeping a sphere over all edges and faces. From a mathematical viewpoint, these block shapes can be seen as the Minkowski sum of a polyhedron and a sphere [BER 03].

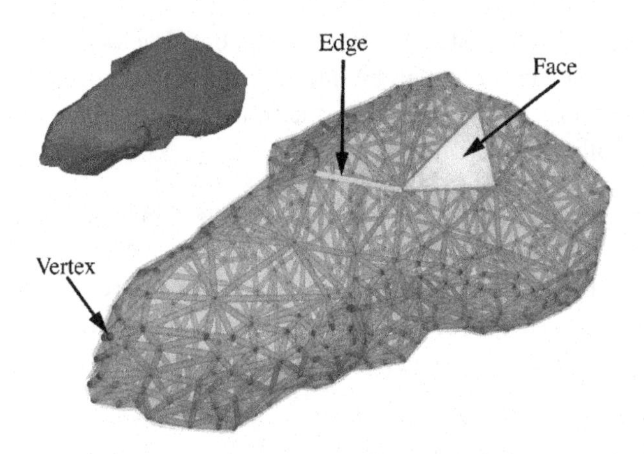

Figure 1.1. *Definition of a sphero-polyhedron illustrated in the case of a non-convex pebble. The vextexes are spheres, the edges are cylinders and the faces are thick 3D polygons (triangles in this picture)*

SP offer several advantages including highly simplified contact detection (i.e. finding contact locations and associated local frameworks) [ALO 08]. Indeed, all the contact configurations between two SP can be reduced to a set of only four types of elementary contacts configurations: *vertex–vertex*, *vertex–edge*, *vertex–face* and *edge–edge*. One can better appreciate the benefit of this approach when considering, for instance, the face–face intersection test: the latter can simply be replaced by a set of edge–edge and vertex–face intersection tests. Among the many other benefits, the SP approach allows the shapes to be concave and/or hollow. Also, the normal vectors at contact are defined without ambiguity.

1.2.2. *Mass properties*

The mass properties of the SP have to be precomputed in order to integrate their dynamic motions. They are the center of mass, the mass (or volume) and the inertia matrix expressed in the principal frame. To estimate all these properties for blocks that may have any geometries (concave, convex, hollow), Monte Carlo (MC) numerical integrations are performed (see [PRE 07]). The basic premise of the method relies on the approximation of the integral of a function f on a volume V:

$$\int_V f \mathrm{d}V \simeq V\langle f\rangle \pm \sqrt{\frac{\langle f^2\rangle - \langle f\rangle^2}{N}} \qquad [1.5]$$

where the symbol $\langle \ldots \rangle$ corresponds to an arithmetic mean of a sample of N points.

The integration procedure is conducted by first setting an *axis-aligned bounding box* (AABB) that closely wraps the body. A set of positions within this AABB is randomly generated by means of a Sobol sequence (for a faster MC integration). To assess whether or not this point $\mathbf{x}$ stands inside the SP, a function $\varphi(\mathbf{x})$ is defined so that it is 1 when the point is inside the shape volume V, and 0 otherwise. In practice, this is verified quite trivially at the vertexes (sphere), at the edges (cylinder) and at the faces (thick three-dimensional [3D] polygon). Then, on the inside polyhedron (i.e. without the Minkowski radius) this is verified with an algorithm based on the oddness of the amount intersection between a semiinfinite ray (starting from the point) and each face. Finally, in more formal terms:

$$\varphi(\mathbf{x}) = \left(\sum_{\text{vertexes}} \varphi_{\text{sphere}}(\mathbf{x})\right) + \left(\sum_{\text{edges}} \varphi_{\text{cylinder}}(\mathbf{x})\right) + \left(\sum_{\text{faces}} \varphi_{\text{polygon}}(\mathbf{x})\right) + \varphi_{\text{polyhedron}}(\mathbf{x}) \qquad [1.6]$$

It then becomes easy to numerically integrate any quantity on V^{AABB} volume using equation [1.5]. The volume of the body is first estimate:

$$V = \int_{V^{\text{AABB}}} \varphi(\mathbf{x})\text{d}V \simeq \langle \varphi(\mathbf{x}) \rangle V^{\text{AABB}} \qquad [1.7]$$

By assuming a volume density ρ uniformly distributed, the mass of the body is $m = \rho V$, and its inertial center $\mathbf{x}_G$ can be obtained by means of an MC integration:

$$\mathbf{x}_G = \frac{1}{V} \int_{V^{\text{AABB}}} \mathbf{x}\, \varphi(\mathbf{x})\text{d}V \simeq \frac{\langle \mathbf{x}\, \varphi(\mathbf{x}) \rangle V^{\text{AABB}}}{V} \qquad [1.8]$$

The symmetric matrix of inertia relative to the point $\mathbf{x}_G$ is also computed using MC integrations for each of the six components:

$$\begin{aligned}
I_{xx}(\mathbf{x}_G) &= \gamma \left\langle \varphi(\mathbf{x}) \left(\delta y^2 + \delta z^2\right) \right\rangle \\
I_{yy}(\mathbf{x}_G) &= \gamma \left\langle \varphi(\mathbf{x}) \left(\delta x^2 + \delta z^2\right) \right\rangle \\
I_{zz}(\mathbf{x}_G) &= \gamma \left\langle \varphi(\mathbf{x}) \left(\delta x^2 + \delta y^2\right) \right\rangle \\
I_{xy}(\mathbf{x}_G) &= -\gamma \left\langle \varphi(\mathbf{x})\, \delta x \delta y \right\rangle \\
I_{xz}(\mathbf{x}_G) &= -\gamma \left\langle \varphi(\mathbf{x})\, \delta x \delta z \right\rangle \\
I_{yz}(\mathbf{x}_G) &= -\gamma \left\langle \varphi(\mathbf{x})\, \delta y \delta z \right\rangle
\end{aligned} \qquad [1.9]$$

where ${}^T(\delta x,\, \delta y,\, \delta z) = (\mathbf{x} - \mathbf{x}_G)$ and the common prefactor is

$$\gamma = \frac{V^{\text{AABB}}}{V} \qquad [1.10]$$

To save memory and computing time, only the eigenvalues extracted from the inertia matrix will be stored as I_1^*/m, I_2^*/m and I_3^*/m. The vertex positions are expressed in the body framework given by the eigenvectors of the inertia matrix, with the origin placed at the mass center of the SP. The position and orientation of the latter is thus defined.

The use of regular shape is sometimes required (small-scale laboratory experiments, characterization of the influence of the shape, etc.). It is in this case advantageous to have a library of shapes with their inertia properties (inertia and volume) precalculated either by hand or by numerical integration depending on the complexity of the form. Such a library was added to the code DEMbox by setting each form in its own framework with chosen dimensions. The scaling is done through a scaling factor H. For example, the positions of the elements constituting a cube are defined relative to the center of the cube, the main axes being those of the edges having a unit length. If a cube of 1.8 cm square has to be taken from the library, a scaling factor $H = 0.18$ will be used (the unit of length being the meter). The volume will be scaled by multiplying the "unit volume" by H^3 and the eigenvalues of inertia (divided by the body mass) by H^2.

1.2.3. *Block motions*

Since an SP is a rigid body, only the time evolution of the mass center position and overall rotation is computed. The movement of the entities that compose the SP (namely the slave bodies) is governed by the relations of rigid motion.

The algorithm for the classical DEM involves two stages for each rigid body i: (1) the computation of the resultant forces $\mathbf{F}_i$ and moment $\mathbf{M}_i$ from volume and contact forces (see section 1.2.6); and (2) the time integration of Newton's second law (for translations) and Euler's equations (for rotations). This movement integration is performed by means of the velocity Verlet scheme [ALL 89], which is a satisfactory compromise between the accuracy of the block velocities (for both translations and rotations) and memory saving.

Each body i is kinematically defined in the global framework by its mass-center position $\mathbf{x}_i$, its mass-center velocity $\mathbf{v}_i$, its angular position given by a unit quaternion $\breve{Q}_i$ and its angular velocity $\boldsymbol{\Omega}_i$. Unit quaternions provide a convenient mathematical notation for representing orientations and rotations of objects in three dimensions. Compared to Euler angles, they are simpler to compose and may avoid the problem of gimbal lock. Compared to rotation matrices, they are more numerically stable and use less memory. Somehow, unit quaternions can be interpreted as rotation matrices but they only hold one scalar value and one vector: $\breve{\boldsymbol{Q}}_i \equiv [s_i, \ \mathbf{V}_i]$. There exists a largely developed mathematical background concerning quaternions, but all this knowledge is not absolutely necessary when dealing with rotations.

For translation motion, the velocity Verlet algorithm can be summarized as follows for each time step Δt:

$$\begin{cases} \mathbf{x}_i(t+\Delta t) = \mathbf{x}_i(t) + \mathbf{v}_i(t)\Delta t + \frac{1}{2}\mathbf{a}_i(t)\Delta t^2 \\ \mathbf{v}_i(t+\Delta t) = \mathbf{v}_i(t) + \frac{1}{2}\left[\mathbf{a}_i(t) + \mathbf{a}_i(t+\Delta t)\right]\Delta t \end{cases} \qquad [1.11]$$

with

$$\mathbf{a}_i = \frac{\mathbf{F}_i}{m_i} - g\mathbf{z}, \qquad [1.12]$$

where m_i is the mass of the body i and $g\mathbf{z}$ is the gravity acceleration.

For rotation motion, the velocity Verlet algorithm is also used to determine the angular positions and velocities of the bodies. It formally reads:

$$\begin{cases} \breve{\boldsymbol{Q}}_i(t+\Delta t) = \breve{\boldsymbol{Q}}_i(t) + \dot{\breve{\boldsymbol{Q}}}_i(t)\Delta t + \frac{1}{2}\ddot{\breve{\boldsymbol{Q}}}_i(t)\Delta t^2 \\ \boldsymbol{\Omega}_i(t+\Delta t) = \boldsymbol{\Omega}_i(t) + \frac{1}{2}\left[\dot{\boldsymbol{\Omega}}_i(t) + \dot{\boldsymbol{\Omega}}_i(t+\Delta t)\right]\Delta t \end{cases} \qquad [1.13]$$

In this scheme, the first and second time derivative of the quaternion can be expressed in terms of the angular velocity vector as:

$$\dot{\breve{\boldsymbol{Q}}}_i(t) = \frac{1}{2}\breve{\boldsymbol{\Omega}}_i(t) \star \breve{\boldsymbol{Q}}_i(t) \qquad [1.14]$$

and

$$\ddot{\breve{\boldsymbol{Q}}}_i(t) = \frac{1}{2}\dot{\breve{\boldsymbol{\Omega}}}_i(t) \star \breve{\boldsymbol{Q}}_i(t) + \frac{1}{4}\breve{\boldsymbol{\Omega}}_i(t) \star \breve{\boldsymbol{\Omega}}_i(t) \star \breve{\boldsymbol{Q}}_i(t) \qquad [1.15]$$

where $\breve{\boldsymbol{\Omega}}$ denotes for the quaternion $[0, \ \boldsymbol{\Omega}]$, and operator $\star$ is the Hamilton product defined by

$$[s_1, \ \mathbf{i}_1] \star [s_2, \ \mathbf{i}_2] = [s_1 s_2 - \mathbf{i}_1\mathbf{i}_2, \ s_1\mathbf{i}_2 + s_2\mathbf{i}_1 + \mathbf{i}_1 \times \mathbf{i}_2] \qquad [1.16]$$

By introducing the relations [1.14] to [1.16] in equation [1.13], an expanded derivation of the rotation noted $\breve{Q}_i(t) \equiv [s_Q(t),\ \mathbf{Q}(t)]$ reads:

$$s_Q(t+\Delta t) = s_Q(t) - \left(\frac{3}{4}\boldsymbol{\Omega}(t)\cdot\mathbf{Q}(t) + \frac{\boldsymbol{\Omega}^2(t)}{8}\right)\Delta t \qquad [1.17]$$

and

$$\mathbf{Q}(t+\Delta t) = \mathbf{Q}(t) + \frac{3}{4}\Big(s_Q(t)\boldsymbol{\Omega}(t) + \boldsymbol{\Omega}(t)\times\mathbf{Q}(t)\Big)\Delta t$$
$$-\frac{\boldsymbol{\Omega}^2(t)}{4}\mathbf{Q}(t) \qquad [1.18]$$

In equation [1.13], the derivative of angular velocities of each body i is obtained from Euler's equations as follows (subscripts i are removed to facilitate reading):

$$\begin{cases} \dot{\Omega}_1^* = \Big(M_1^* - (I_3^* - I_2^*)\Omega_2^*\Omega_3^*\Big)/I_1^* \\ \dot{\Omega}_2^* = \Big(M_2^* - (I_1^* - I_3^*)\Omega_3^*\Omega_1^*\Big)/I_2^* \\ \dot{\Omega}_3^* = \Big(M_3^* - (I_2^* - I_1^*)\Omega_1^*\Omega_2^*\Big)/I_3^* \end{cases} \qquad [1.19]$$

where components 1, 2 and 3 are expressed in the body frame (superscripts $*$ are added here for memory). To express the resultant moment $\mathbf{M}_G$ and the angular velocity $\boldsymbol{\Omega}$ in the framework of a body i, a rotation corresponding to the inverse of the block orientation must be applied. This comes down to build and use a rotation matrix from the conjugate of the orientation quaternion, i.e. with the imaginary part having the opposite sign:

$$\mathbf{M}_G \overset{\mathrm{conj}(\breve{Q})}{\longrightarrow} \mathbf{M}_G^* \quad \text{and} \quad \boldsymbol{\Omega} \overset{\mathrm{conj}(\breve{Q})}{\longrightarrow} \boldsymbol{\Omega}^* \qquad [1.20]$$

The components of the angular acceleration vector in the global framework is obtained by rotating the one from equation [1.19]:

$$\dot{\boldsymbol{\Omega}}^* \overset{\breve{Q}}{\longrightarrow} \dot{\boldsymbol{\Omega}} \qquad [1.21]$$

It might be important to note that, in equations [1.12] and [1.19], $\mathbf{F}_i$ and $\mathbf{M}_i$ are obtained from the contact force laws, which depend on body positions ($\mathbf{r}$ and $\hat{q}$) at time t and their mean velocities between times $t - \Delta t$ and t. In order to synthesize the algorithm used, the following pseudo-code specifies the steps of calculation.

1: **if** (update of neighborhood is required) **then**
2: Update body neighborhoods (section 1.2.7)
3: **end if**
4: **if** (data storing is mandatory) **then**
5: **Save** data
6: **end if**
7: **Update** body positions and orientations
8: **Compute** velocities "at the middle of the time increment" (i.e. equations [1.11] and [1.13] without the acceleration terms at time $t + \Delta t$)
9: **Compute** the interaction forces and moments (section 1.2.6)
10: **Update** the velocities by accounting for the acceleration terms resulting from the interaction forces (and moments)

1.2.4. *Pre-existing discontinuities*

Discontinuities of a rock mass are related to the phenomena occurring during their formation and the stress states that it has undergone during its geological history. It is therefore natural that they are not randomly oriented; instead, they are organized in families with the same orientation and same characteristics. These families are usually highlighted by representing, in stereographic projection, normal to the surface discontinuities observed in a homogeneous area.

A numerical procedure for cutting a three-dimensional SP volume according to different families of discontinuities has been developed. The core of this procedure, which is purely geometric, is not described in much detail here. Let just say that it consists of splitting a volume of any shape according to several plane families successively. Figure 1.2 provides an illustration of the principle of the procedure implemented in a module DEMbox. It is important to keep in mind that a family of discontinuities is defined by (1) a normal vector common to all discontinuity planes, (2) a point belonging to the first plane, (3) a point belonging to the last plane and (4) the distance (chosen constant) between the planes. The extent of the planes of discontinuity are infinite, which implies that each cut passes through the mass from one end to the other, which is not necessary the case with *discrete fracture models*.

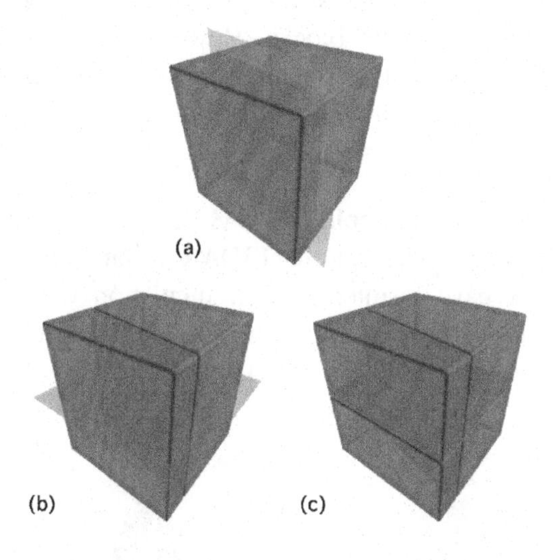

Figure 1.2. *A schematic illustration of the cutting procedure along families of discontinuity planes. The initial volume, represented here as a cube, is a sphero-polyhedron: a) it is first cut along the first family of discontinuities (only a single yellow plane is shown); b) each resulting block will be cut along the following family of discontinuities (only a single green plane is shown); c) when all the cuts are made, the result is an assembly of blocks that will be able to interact when they will be destabilized by gravity*

This procedure gives rise to realistic geometries of blocks. They are issued from a 3D digital model of the unstable volume (in the standard STL format for example) and a detailed knowledge of the fracturing resulting from a site investigation or from geological databases. Some applications to real sites are shown in Chapter 4.

1.2.5. *Digital terrain model*

The topology of the terrain can be assessed by means of different techniques (aerial LIDAR scans, stereocorrelation of photographic images, etc.). Such a scan of the terrain generally results, after some postprocessing, in a triangular mesh forming the *digital surface model* (DSM) of the terrain. From the point of view of DEMbox model, each triangle is a spherotriangle having all its degrees of freedom blocked. All the advantages of SP still exist for both the terrain topology and its interactions with the moving blocks. However, the question of the sensitivity to the size resolution (related somehow to the bumpyness of the terrain) remains, even if the use of Minkowski radii smooths the surface.

Figure 1.3 shows an example of DSM. In this example, the relief has been scanned by means of the LIDAR technique, and different digital treatments have resulted in a triangulation with the desired fineness. The inset in the top-right corner illustrates how the spherotriangles are superimposed onto the DSM.

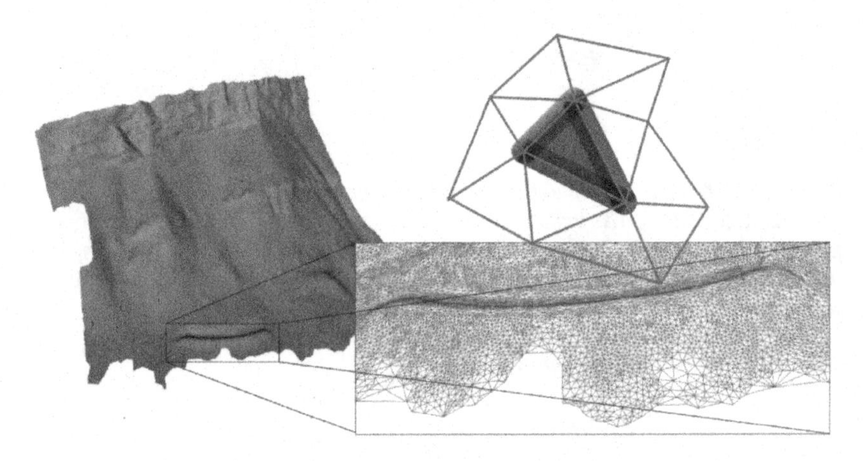

Figure 1.3. *Digital surface model used to model the spread of a rock avalanche triggered artificially as part of a preventive purge of unstable rocky shell of Neron mount near Grenoble (France) in 2011 Image from [BOT 14]*

1.2.6. *Contact force laws*

The goodness of a model can be evaluated from its ability to reflect concrete physical phenomena, but also by the easiness with which the required parameters are accessible. In other words, the current stakes in modeling gravity hazards concern not only the development of powerful numerical tools capable of simulating complex rheologies but also the ability of an operator to identify and assign model parameters. With this in mind, a dissipative model of contact–collision is depicted. The model is intentionally minimalist. For rockfall modeling, it is believed that it includes the minimum number of features necessary to take account of (1) the deformability of the contact zones, (2) a dissipation related to normal collisions and (3) a dissipation associated with tangential relative displacements.

The model depicted here is based on body shapes that are more realistic than spheres, as seen previously. This is actually a crucial feature of the model. The other crucial feature concerns collision (or contact) force laws. A simple formulation has been used; nevertheless, it incorporates the energy dissipation due to block collisions. Considering the huge amount of uncertainties related to a natural event, it seemed totally impossible to predict the exact trajectory of each particle in the flow. Thus, it was decided to focus on the energy loss associated with each impact, rather than to reproduce in smallest details the exact physical phenomena related to this impact. The energy loss may result from very complex physical mechanisms (heat production, wave propagation, microcracking, block chipping, etc.) for which it is reasonable to admit that they are beyond understanding of collective behavior of the blocks in a flow. Moreover, the local mechanisms do not need to be precisely identified, especially since it will be necessary to identify the parameters involved. Minimalist laws were opted for, where only the rate of energy loss and the friction are required to dissipate the kinematic energy of the blocks. In other words, a coarser scale has been considered to take into account force transmission and dissipation mechanisms in a granular assembly that flows and then stops. It is important to stress here that this coarser scale is not non-physical but ignores some physical mechanisms that are involved at smaller scales. It will be seen in the following chapters that the laws proposed here are sufficient to satisfactorily describe the main rebound patterns – even the most erratic – provided that the minimum time resolution of interest is longer than a collision duration. Obviously, the model here should not be used when dealing with some other features for which the force evolution during the contact/collision, such acoustic wave propagation in confined granular systems [SOM 05] (anyway, it is not why it is designed for).

Rock avalanches involve dynamic block movements. For this reason, damping models, which affect block movements with an artificial parachute, cannot be used since it would lead to non-physical behavior. Another solution is to account for a local viscous damping at a contact level. This solution was also rejected because, although it introduces a viscosity parameter that can be connected to a well-defined dissipation rate in the case of single contact [TSU 92], it is ill-defined in the particular case of the multiple contacts involved in the interactions of SP. More precisely, the effective mass m_{eff} involved in the critical viscosity $2\sqrt{m_{\text{eff}} k_n}$ is not well defined for complex shapes and should depend on the positions of the contact points and their number.

1.2.6.1. *Normal force*

The simplest formulation for the normal force f_n makes use of a linear elastic law with two different stiffnesses in the case of loading or unloading (respectively, k_n^+ and k_n^-) [BAN 09]. When the overlap h_n increases (i.e. $\Delta h_n \geq 0$), the normal force increment reads:

$$\Delta f_n = k_n^+ \, \Delta h_n \qquad [1.22]$$

Otherwise, if $\Delta h_n < 0$ and $h_n > 0$, the force f_n is given by:

$$f_n = k_n^- \, h_n \qquad [1.23]$$

Figure 1.4(a) is a plot of this force law, which illustrates the role of incremental loading. It should be noted that the literature provides other contact laws that also introduce energy dissipation by means of a difference of normal stiffness for loading and unloading (see [LUD 03, OGE 98]). These laws have all the desired properties for rock flow simulations and provide accurate results for the impacts. Unfortunately, this leads to numerical issues at the end of the flow where persistent contacts tend to "oscillate", and an additional damping (or smaller time step) is required. On the contrary, the contact law shown in Figure 1.4(a) has the ability to naturally dissipate the energy during these oscillations. It is thus much more convenient for the numerical modeling of rock flows under consideration here. It has the added advantage of using a single parameter for normal energy dissipation without the need to postulate on all phenomena actually involved (e.g. viscosity, plasticity). Note that nothing prevents us from defining e_n^2 as a function of the collision velocity. Since the focus is not on the actual value of the repulsive force, the "force jump" used in the proposed contact law is not an issue. This sounds non-physical from a contact mechanics point of view, but the apparent lack of realism in the definition of contact force laws is irrelevant for the problem treated here.

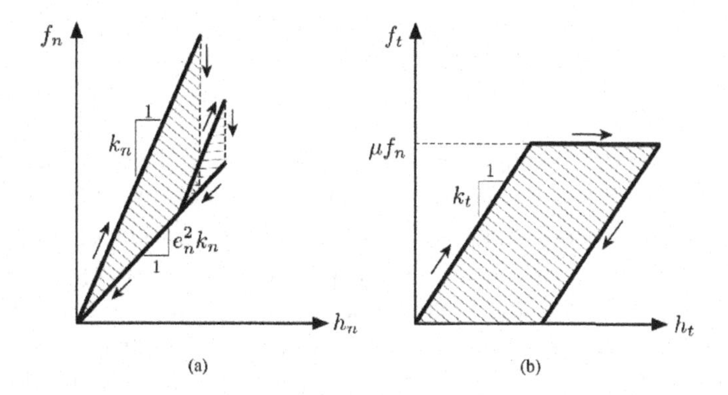

Figure 1.4. *A schematic representation of the force laws: a) normal contact force as a function of the overlap; b) tangent friction force as a function of the sliding displacement*

Now consider a normal collision, that is – in this case of smooth contact formulation – a contact loading up to a given overlap $h_n^{\max}$, immediately followed by an unloading until $f_n = 0$. After the impact, one part $k_n^-(h_n^{\max})^2/2$ has been restored, and since the maximum energy that can be restored is $k_n^+(h_n^{\max})^2/2$, the dissipation rate in the normal direction is:

$$e_n^2 = k_n^-/k_n^+ \qquad [1.24]$$

In this definition, the square indicates that the dissipation rate is expressed in terms of an energy rate. In this way, the dissipation rate can also be defined as the ratio between the relative normal velocity before and after the collision:

$$e_n = v_n^-/v_n^+ \qquad [1.25]$$

1.2.6.2. *Tangential force*

The tangential force $\mathbf{f}_t$ stands in the plane perpendicular to the contact normal and it is in the direction opposite to the sliding direction. For hard contacts (i.e. high values of k_n), this force obeys a Coulomb friction law:

$$\mathbf{f}_t = \min\left\{\mathbf{f}_t^{\text{trial}}\,;\, \mu f_n \frac{\mathbf{f}_t^{\text{trial}}}{|\mathbf{f}_t^{\text{trial}}|}\right\} \qquad [1.26]$$

where μ is the coefficient of friction and $\mathbf{f}_t^{\text{trial}}$ is a trial elastic force that may return back to a thresholded magnitude proportional to the normal force f_n. For smooth contacts, the same law applies but the physical meaning of μ is, this time, slightly different: it incorporates an abutment force. For this reason, the coefficient μ is better named *tangential coefficient of dissipation* in order to highlight the fact that the tangent force is not necessary issued from Coulombic friction in the context of smooth collision.

Coming back to equation [1.26], the trial force $\mathbf{f}_t^{\text{trial}}$ is incrementally updated as a function of the increment of relative displacement $\Delta\mathbf{f}_t^{\text{trial}} = k_t \Delta\mathbf{h}_t$ in the sliding direction at each time step – k_t is the tangential elastic stiffness. However, during the sliding, the contact normal may rotate, and this rotation is accounted for by the strategy developed in [HAR 88]:

$$\mathbf{f}_t^{\text{test}} = c_2\Big(c_1(\mathbf{f}_t)\Big) - k_t\Delta\mathbf{h}_t \qquad [1.27]$$

where

$$c_1 : \mathbf{f}_t \mapsto \mathbf{f}_t - \mathbf{f}_t \times \Big(\mathbf{n}(t) \times \mathbf{n}(t+\Delta t)\Big) \qquad [1.28]$$

and

$$c_2 : \mathbf{f}_t \mapsto \mathbf{f}_t - \mathbf{f}_t \times \left(\frac{\Delta t}{2}(\mathbf{\Omega}_i + \mathbf{\Omega}_j)\cdot \mathbf{n}^2(t+\Delta t)\right) \qquad [1.29]$$

A two-dimensional picture of this law is shown in Figure 1.4(b) assuming a constant normal force.

1.2.6.3. *Rolling resistance*

The propagation slopes and stop areas of rock avalanches are sometimes very steep. It may not be possible, in this case, to stop a block. To be convinced of that, Figure 1.5 shows two situations where, for a given inclination α of the slope, a bloc will or will not be able to stop. Intuitively, it is understandable that a flat block may have the ability to stop on its largest face, even if the slope is steep, while a cubic block may not necessarily be able to.

More pragmatically, a necessary condition for a given form of block can be stopped on an inclined slope is the position of the center of mass does not exceed the vertical passing through the point of contact. The form of a block can be characterized by an angle β defined as the maximum angle between the normal to a face and the vector connecting the center of mass at the vertexes of that face (see Figure 1.5(a)). The necessary condition for stabilization can then formally be expressed as

$$\alpha - \beta < 0 \qquad [1.30]$$

In other words, if this condition is not satisfied, the block will not be able to stop, unless a moment at the contact point resists the lever action of the weight. This moment may physically arise from the indentation of the block in a soft soil. Analogously to the friction law, the rolling resistance formally reads:

$$\mathbf{C} = \min(\mathbf{C}^{\text{test}};\ \mu_r \ell f_n \frac{\mathbf{C}^{\text{test}}}{|\mathbf{C}^{\text{test}}|}) \quad \text{with}$$

$$\mathbf{C}^{\text{test}} = \mathbf{C} - k_r(\Omega_j - \Omega_i)\Delta t \qquad [1.31]$$

where k_r is a parameter of elastic stiffness in rotation, ℓ is the distance between the center of mass and the point of contact and μ_r is a rolling dissipation coefficient. For the equilibrium of the block in the situation in Figure 1.5(b), it is necessary that the moment due to the weight of the block relative to the contact point $\ell \sin(\alpha - \beta)mg$ be outweighed by the resisting moment limited to a minimum value $\mu_r \ell f_n = \mu_r \ell mg \cos(\alpha)$. This results in a minimum value $\mu_r^{\min}$ of magnitude of the rolling resistance coefficient for that stabilization is possible:

$$\mu_r^{\min} = \frac{\sin(\alpha - \beta)}{\cos(\alpha)} \qquad [1.32]$$

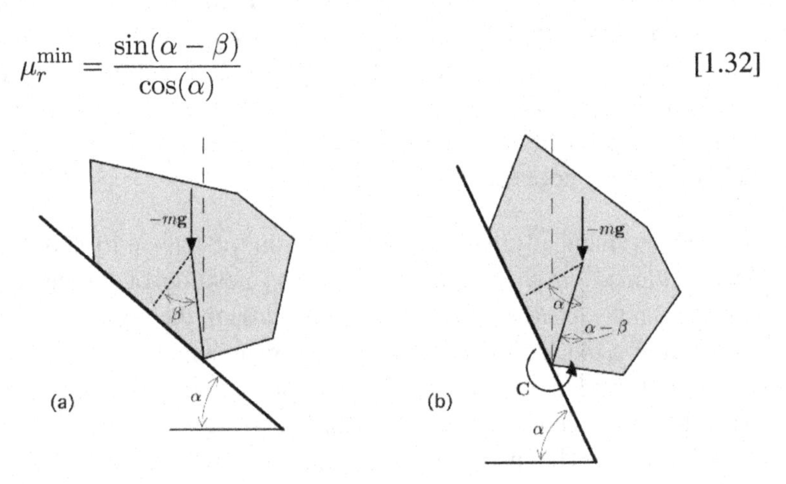

Figure 1.5. *Schematic representation of a) a stable position and b) a position potentially unstable*

Below this value, a block having a form characterized by β will not stop on a slope of inclination α. For example, a cube can be characterized by $\beta = \pi/4$, and can stop on a slope inclined at $\alpha = \pi/3$ if the rolling resistance coefficient μ_r exceeds $2\sin(\pi/12) \simeq 9.2 \times 10^{-3}$ (in addition to the tangential force).

1.2.7. *Neighborhood of each block*

The purpose of the procedures described in this section is to build a list of potential contacts – that is a neighbor list – in the most efficient way. In many situations, the update of this list has a significant cost to the overall computational time (although its role is precisely to optimize the calculations). The most basic method to build a neighbor list involves testing the vicinity of all possible pairs of blocks, which leads to a number of operations equal to the square of the number of elements. In terms of Big-O notation, this is referred to as an algorithm with $\mathcal{O}(N^2)$ complexity; however, insofar as the updating of the list can be done with a time interval greater than the time step, a modest gain is still achieved.

For the problem of interest – involving complex shapes together with simulated duration in the order of a few minutes – the effective update of the neighbor list is a critical aspect. This is why a special effort needs to be devoted to the adaptation of more sophisticated strategies, often borrowed from the world of 3D video games. These strategies focus, on the one hand, on the volumes surrounding the blocks, and on the other hand, on the algorithm itself by targeting a $\mathcal{O}(N)$ complexity.

SP are, as already explained at the beginning of the chapter, composed of convex subelements (spheres, cylinders and thick 3D-polygons). Stating whether two SP are close would require a test for each pair of subelement they are constituted, which would involve a large number of operations. The logical strategy with this type of object is to test – with a certain frequency – the proximity of volumes that surround each SP (bounding shapes), then to test the proximity between the subelements of the SP identified as being close with a frequency that may be different. In `DEMbox`, two distances D_v and d_v

are defined. Below the distance D_v, two SP are close enough for the proximity of the subelements to be tested in turn, and added to a list of neighbors when there distance is lower or equal to d_v. Each of these distances is associated with a period of update N_D and N_d expressed as a number of time steps. For consistency, it is necessary that N_d is a multiple of N_D. It is quite possible to define a strategy for the dynamic assignment of distance values and updating periods, but for the sake of simplicity and efficiency, this solution was chosen to not be implemented. A more global solution, based on an assumed maximum velocity, was used to define them.

The choice of a bounding shape has a significant influence on the reconstruction efficiency, especially if the objects have an elongated shape. Both AABB and oriented bounding box (OBB) have been tested. For the latter, a geometric algorithm was developed and used to determine the box with the smallest volume. This algorithm is not described here, but note that the axes of the optimum OBB does not necessarily correspond to the inertia axes of the SP. Testing the proximity of two OBB is slightly more expensive than testing the proximity of two AABB; however, for purely geometrical reasons, the use of OBBs is actually more efficient because it limits the scope of objects in all directions as the shown in Figure 1.6. Thus, with dense flow simulations, seven times shorter durations were obtained when OBBs were used. The algorithm performance obviously depends on the size of cells relative to the typical block size. Therefore, there is an optimum value of this ratio which must have – according to our own experience – a value in the order of 2. This value can vary with the grains size dispersion and the elongation of the blocks.

All strategies so far described enable a significant reduction in the time required for updating the neighbor list, but the complexity of the algorithm remains $\mathcal{O}(N^2)$. To overcome this, the strategy of "link-cells" has been implemented. It consists of dividing the entire system (i.e. the AABB that surrounds all the elements of the simulation) into structured cells (a grid) associated with a list of elements it contains, and a dummy cell associated with the elements that are too large to be contained in the structured cell, such as the infinite planes. For a given object, the proximity test is achieved only with the other objects in its own cell, as well as those located in neighboring cells. The number of involved cells for the search of neighbors of an object in 3D is *constant* and equals 27. Figure 1.7 gives a 2D representation of the cells involved in the search of neighbors of an element; in this case, the cells tested are nine in number. Considering that the number of elements per cell is limited, the number of operations per elements is nearly constant and the complexity of the algorithm is in the worst case limited to $\mathcal{O}(N)$.

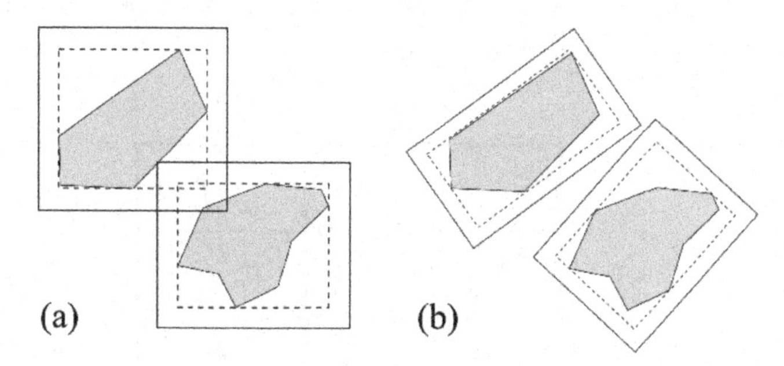

Figure 1.6. *A 2D illustration of the bounding boxes AABB a) and OBB b). Dashed boxes are the tightest boxes, and the solid line boxes are the same boxes dilated with a distance D_v (or d_v for subelements). When two solid-line boxes overlap, the sphero-pohedra are stated to be "close". This image shows that, even if the relative positions of the objects and the dilatation distances are the same, they are wrongly stated to be "close" with AABBs, while they are not with OBBs*

An added improvement is to limit the link cells to the moving blocks instead of the entire system, which include the terrain elements. Actually this latter elements are many and never interact the ones with the others. This small change is capable of inducing a gain of the order of three to four in some of the simulations.

1.3. Material point method

In the previous section, the DEM with complex block shapes has been introduced. This method is able to model the granular scale physics of bodies that interact. In return, the number of blocks is very limited because the computation time can quickly become prohibitive. In a context of rock avalanche engineering, a more macroscopic vision of the flowing mass has to be adopted when the amount of blocks involved is "too large". In other words, a continuum approach needs to be used. But this approach has to deal with plasticity and very large strains. A kind of description which is both Eulerian and Lagrangian is thus required to avoid the complex challenge of remeshing. Among the mesh-free approaches, we chose to use the material point method (MPM), which is a relatively new technique that brings a lot of benefits – but also some drawbacks.

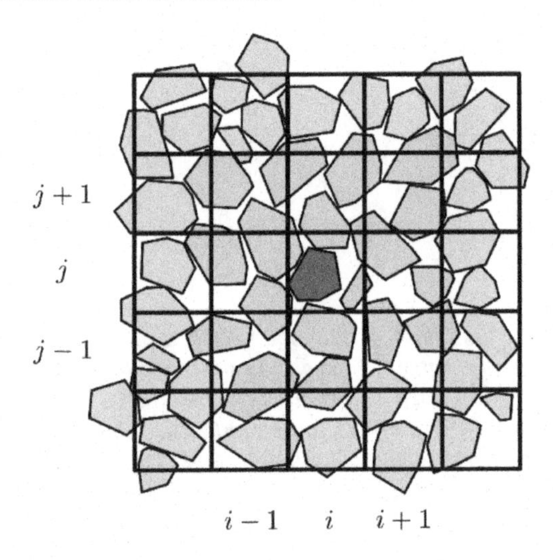

Figure 1.7. *The link-cell strategy: the proximity test of the red element in the cell (i, j) is performed only with the green elements belonging to the nine cells ranging from $(i-1, j-1)$ to $(i+1, j+1)$*

As one of the most straightforward spatial discretization methods, the MPM is an extension to solid mechanics problems of a hydrodynamics code names fluid-implicit particle (FLIP), which, in turn, evolved from the particle-in-cell method dating back to the pioneering work of Harlow [HAR 64]. The motivation of the development was to simulate problems such as impact/contact, penetration and perforation with history-dependent internal state variables, as advocated in several important publications about the MPM [SUL 94, SUL 95, SUL 96]. The essential idea is to take advantage of both the Eulerian and Lagrangian methods.

The MPM is a numerical method for the solution of problems in continuum mechanics, in which large deformations can be accomplished. This method solves the variational form of the conservation of momentum by discretizing the continuum into MPs. Since the mass remains unchanged during the computation, the conservation of mass is also satisfied. A background grid is then used to project the information carried by the MPs and solve the equations of motion. The solutions found in the mesh are used to update the information contained in the MPs. This interpolation/extrapolation between the mesh and the MPs is done using shape functions.

In what follows, an attempt is made to provide the unfamiliar reader with a basic introduction to the MPM implemented in a C++ code named `MPMbox`. Only the essential basics of the method will be addressed. The equations are given here for the 2D case, but switching to 3D is effortless.

1.3.1. *Conservation equations*

By considering a medium described as a continuum within the domain Ω, the conservation of mass is governed by the continuity equation:

$$\frac{\partial \rho(\mathbf{x},t)}{\partial t} + \nabla \cdot \Big(\rho(\mathbf{x},t)\ \mathbf{v}(\mathbf{x},t)\Big) = 0 \qquad [1.33]$$

where ρ is the material density and $\mathbf{v}$ is the velocity at position $\mathbf{x}$ and time t. The symbol ∇ denotes for the gradient operator. In the context of the infinitesimal strain theory, the conservation of linear momentum is defined as

$$\rho(\mathbf{x},t)\ \mathbf{a}(\mathbf{x},t) = \nabla \cdot \boldsymbol{\sigma}(\mathbf{x},t) + \mathbf{b}(\mathbf{x},t) \qquad [1.34]$$

where $\boldsymbol{\sigma}$ is the Cauchy stress tensor, $\mathbf{b}$ represents the body force and $\mathbf{a}$ is the acceleration. The conservation of angular momentum is ensured by the symmetry of the Cauchy stress. The continuity equation [1.33] and momentum equation [1.34] must be supplemented with a suitable constitutive equation. This question will be addressed in section 1.3.5.

The conservation of the linear momentum can be rewritten into a weak formulation, just like the finite element method does:

$$\int_{\Omega} \rho \mathbf{a} \cdot \tilde{\mathbf{v}} \mathrm{d}\Omega = -\int_{\Omega} \rho \boldsymbol{\sigma}{:}\nabla \tilde{\mathbf{v}} \mathrm{d}\Omega + \int_{\Omega} \rho \mathbf{b} \cdot \tilde{\mathbf{v}} \mathrm{d}\Omega + \int_{\partial\Omega} \rho \boldsymbol{\tau} \cdot \tilde{\mathbf{v}} \mathrm{d}\partial\Omega \qquad [1.35]$$

where $\tilde{\mathbf{v}}$ is a test function.

1.3.2. *Discretization and continuity relations*

From the weak formulation of the conservation equations, it is possible to derive a set of relations that constitute the core of the method [SUL 94, SUL 95]. Before expressing these relations, let us describe the discretization employed with the MPM sketched in Figure 1.8. The whole domain is divided into a number of pieces with their masses lumped at single positions. These points are the so-called MPs that are sometimes improperly called "particles". This latter term will not be used to avoid confusion with the DEM. The conservation of the total mass is achieved by equation [1.33] and the fact that the mass held by a MP does not vary. The continuity of the domain Ω is provided through the use of a fix grid where the momentum equations are solved. The nodes that constitute the grid overlap both the initial domain Ω and the places where the MPs are supposed to go.

To distinguish between MPs and nodes, the subscripts p and i will, respectively, be used. With this convention, the positions of the MPs are denoted by $\mathbf{x}_p$, and the positions of the nodes are denoted by $\mathbf{x}_i$. At this point, it is interesting to remark that the MPs that transport the state variables of a piece of matter provide a Lagrangian description of the continuum domain. On the other hand, the nodes of the grid provide an Eulerian description.

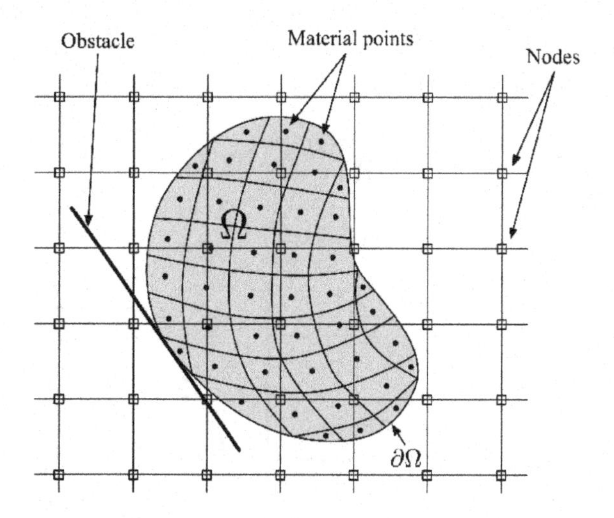

Figure 1.8. *Discretization used with the MPM. Lagrangian MPs (black dots) overlaid over an Eulerian grid of nodes (squared symbols). The gray area denotes the physical domains* Ω *of the material body. The obstacles can interact with the body at its interface* $\partial\Omega$

For simplicity's sake, the fixed grid, which is structured, is made up of regular 4-nodes elements (linear). The elements are numbered from left to right, and from bottom to top as shown in Figure 1.9(a) that also shows the grid characteristics. The local numbering of the nodes starts in the lower-left corner and increases anticlockwise (Figure 1.9(b)). Because of this data structure, identifying the element e to which a MP belongs is straightforward (provided that the MP remains inside the grid):

$$e(\mathbf{x}_p) = \text{trunc}\left(\frac{\mathbf{x}_p \cdot \mathbf{e}_x}{\ell_x}\right) + \text{trunc}\left(\frac{\mathbf{x}_p \cdot \mathbf{e}_y}{\ell_y}\right) n_x \qquad [1.36]$$

The shape functions for a MP standing inside a quad element reads:

$$N_p^I(\mathbf{x}_p) = \left(1 - |\xi^I(\mathbf{x}_p)|\right)\left(1 - |\eta^I(\mathbf{x}_p)|\right) \qquad [1.37]$$

where $\xi^I(\mathbf{x}_p) = (\mathbf{x}_p - \mathbf{x}_I) \cdot \mathbf{e}_x/\ell_x$ and $\eta^I(\mathbf{x}_p) = (\mathbf{x}_p - \mathbf{x}_I) \cdot \mathbf{e}_y/\ell_y$ are the normalized distances relative to node I, which is the local number in the local basis of the quad reference element. The gradient of the shape function can easily be determined by derivation with respect to the global coordinates x and y:

$$\frac{\partial N_p^I(\mathbf{x}_p)}{\partial x} = \frac{1}{\ell_x}\frac{\partial N_p^I(\mathbf{x}_p)}{\partial \xi} \quad \text{and} \quad \frac{\partial N_p^I(\mathbf{x}_p)}{\partial y} = \frac{1}{\ell_y}\frac{\partial N_p^I(\mathbf{x}_p)}{\partial \eta} \qquad [1.38]$$

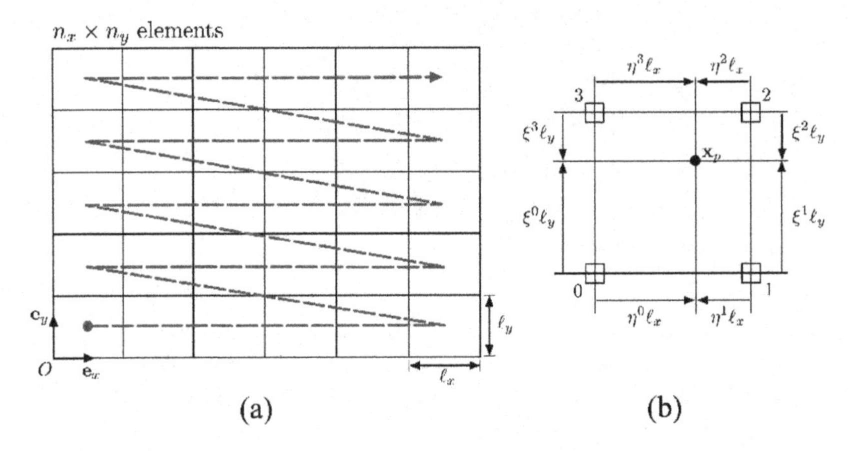

Figure 1.9. *Parameters for the structured grid and the quad elements*

For convenience, the notation $N_{ip} \equiv N_p^I(\mathbf{x}_p)$ is used, where $I = 0$, 1, 2 or 3 as a function of the position of the element e containing the MP relative to the node position: top-right, top-left, bottom-left or bottom-right, respectively.

The purpose of the shape functions is to perform *mapping* from nodes to MPs, or from MPs to nodes as illustrated in Figure 1.10. These mapping procedures, for an arbitrary quantity λ, reads:

$$\lambda_i = \frac{\sum_p \lambda_p N_{ip}}{\sum_p N_{ip}} \quad \text{and} \quad \lambda_p = \sum_i \lambda_i N_{ip} \qquad [1.39]$$

It is important to stress that the node-to-MP mapping is a simple local interpolation, while the MP-to-node mapping realizes an averaging over the support of the shape function. Any quantity held by a MP can thus be smoothed over space by applying consecutively the two mappings.

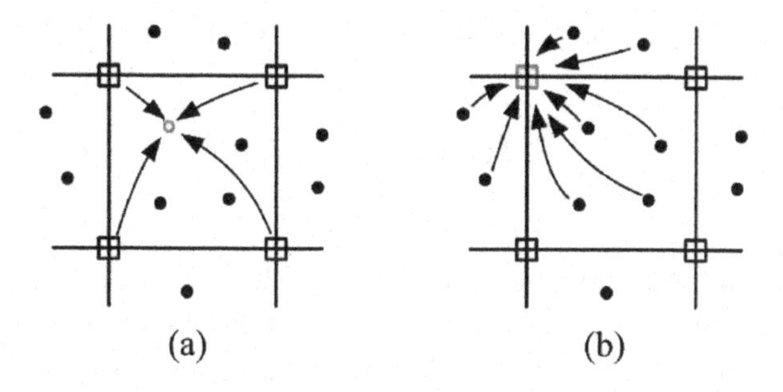

Figure 1.10. *Mapping a) from nodes to a material point and b) from material points toward a node*

1.3.3. *Explicit time integration*

The step-by-step procedure for an explicit time increment begins with the computation of a nodal mass:

$$m_i(t) = \sum_p m_p \, N_{ip}(t) \tag{1.40}$$

These mass transfers from the MPs toward the nodes do not correspond to a mapping as defined earlier, but rather to the definition of the lumped mass.

The MP velocities are then mapped toward the nodes:

$$\mathbf{v}_i(t) = \frac{1}{m_i(t)} \sum_p m_p \, \mathbf{v}_p(t) \, N_{ip}(t) \tag{1.41}$$

Indeed, for numerical considerations, the linear momentum $\mathbf{q}_i = m_i \mathbf{v}_i$ and $\mathbf{q}_p = m_p \mathbf{v}_p$ for, respectively, nodes and MPs is a much more convenient variable to "hold" the velocities.

Considering the constitutive model has been applied (see section 1.3.5), the nodal forces can be computed by mapping the resulting MP accelerations toward the nodes. First, the internal acceleration issued from the local conservation of linear momentum $\mathbf{a}_p^{\text{int}} = -\boldsymbol{\nabla} \cdot \boldsymbol{\sigma}_p / \rho_p$ leads to the following expression of the internal force:

$$\mathbf{f}_i^{\text{int}}(t) = -\sum_p V_p(t) \left(\boldsymbol{\sigma}_p(t) \cdot \boldsymbol{\nabla} N_{ip}(t) \right) \qquad [1.42]$$

where V_p is the MP volume, which needs to be updated by using either the gradient transformation matrix $V_p = \det(\mathbf{F}_p) V_p^0$ or the total strain $V_p = \left(1 + \text{tr}(\boldsymbol{\varepsilon}_p)\right) V_p^0$ depending on what variable is employed by the constitutive model – and thus regularly updated.

The second acceleration term $\mathbf{a}_p^{\text{obs}}$ results from the interaction of the MPs with the obstacles. This term can be computed by using the same force laws than those defined for discrete element simulations (see section 1.2.6). The MP-to-node mapping of these accelerations results in the following expression:

$$\mathbf{f}_i^{\text{obs}}(t) = \sum_p \mathbf{f}_p^{\text{obs}} N_{ip} \qquad [1.43]$$

The third acceleration term is the constant gravity field $\mathbf{g}$; after mapping, the corresponding external nodal force reads:

$$\mathbf{f}_i^{\text{ext}}(t) = \sum_p m_p \mathbf{g} N_{ip} \qquad [1.44]$$

Finally, the motion of the MPs can be obtained by stating the dynamic balance at the node level:

$$m_i(t) \dot{\mathbf{v}}_i(t) = \mathbf{f}_i^{\text{int}}(t) + \mathbf{f}_i^{\text{obs}}(t) + \mathbf{f}_i^{\text{ext}}(t) \qquad [1.45]$$

The nodal accelerations $\dot{\mathbf{v}}_i$ and velocities $\mathbf{v}_i$ are then mapped back toward the MPs in order to perform an explicit time integration of the movement. The following pseudo-code summarizes what happen within a single explicit time-step:

1: **Discard** the previous grid
2: **Compute** the shape functions N_{ip} and their gradients ∇N_{ip} (equations [1.37] and [1.38])
3: **Compute** the lumped nodal mass (equation [1.40])
4: **Map** the velocities $\mathbf{v}_{p \to i}$ (equation [1.41])
5: **if** (Update Stress First) **then**
6: Apply the **constitutive model** to compute $\boldsymbol{\sigma}_p$ (section 1.3.5)
7: **end if**
8: **Compute** node forces f_i^{int}, f_i^{ext} and f_i^{obs} (equations [1.42]–[1.44])
9: **Compute** the nodal accelerations (equation [1.45])
10: **Update** nodal velocities $\mathbf{v}_i \leftarrow \mathbf{v}_i + \dot{\mathbf{v}}_i \Delta t$
11: **Map** $\mathbf{v}_{i \to p}$ and $\dot{\mathbf{v}}_{i \to p}$
12: **Update** MP positions $\mathbf{x}_p \leftarrow \mathbf{x}_p + \mathbf{v}_p \Delta t$
13: **Update** MP velocities $\mathbf{v}_p \leftarrow \mathbf{v}_p + \dot{\mathbf{v}}_p \Delta t$
14: **if** (Update Stress Last) **then**
15: Apply the **constitutive model** to compute $\boldsymbol{\sigma}_p$ (section 1.3.5)
16: **end if**

The update of the stress can be made just before the computation of the internal force (*update stress first* (USF)) or after the motion of MPs (*update stress last* (USL)) [BAR 02]. The USF strategy seems to be provide a better conservation of the energy, although it is not perfect: it slightly overestimates the total energy. Nairm [NAI 03] proposed a *modified USL* that underestimates the total energy. A good compromise can be reached when averaging the two stresses issued from the two strategies; this is the *update stress averaged.*

Extra attention must be paid concerning the nodal masses that can be very close to zero when an MP stands alone within an element and this MP is far from the node. In this case, dividing by the nodal mass can lead to too great number or a "divide-by-zero" error. To overcome this issue, the smallest nodal mass is set to a minimum value ε_m.

1.3.4. *Stability*

1.3.4.1. *Cell crossing instability*

As seen up to now, MPM presents a number of benefits, including its high simplicity from an implementation point of view. Unluckily, it suffers also from a noticeable drawback that, somehow, spoils the approach: the so-called *cell crossing instability*. To understand this phenomenon, let us consider the computation of the internal force (equation [1.42]) that involves the gradients of shape functions. Since the standard version of MPM uses linear shape functions, their space derivatives have a jump at the level of the cell limits. The linear shape functions are said to be C^0 continuous, meaning that they are not continuously derivable. As a result, for a MP that leaves a cell to enter another one, the internal force will also present a jump reflected by a numerical instability. Fortunately, several solutions have been proposed to mitigate the cell crossing instability:

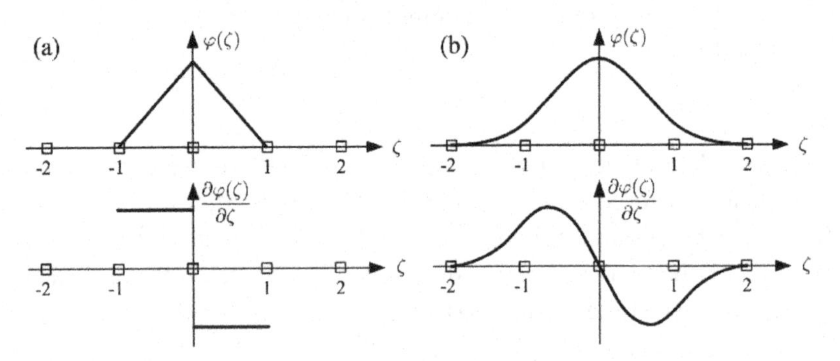

Figure 1.11. *Basis function*

– the first one, which is also the simplest one, is the use of a C^1 (at least) continuous shape function in place of the linear C^0 shape function. In this way, the cell crossing jump is canceled. High-order *B-splines* can be used for this purpose, and this is what `MPMbox` implements. For instance, the one-direction basis function φ_i can be expressed in the local basis around the node i by a cubic B-spline as a function of the local variable ζ (or equivalently for η):

$$\varphi_i(\zeta) = \begin{cases} \frac{1}{2}|\zeta|^3 - \zeta^2 + \frac{2}{3} & \text{if } \ 0 \leq \zeta < 1 \\ -\frac{1}{6}|\zeta|^3 + \zeta^2 - 2|\zeta| + \frac{4}{3} & \text{if } \ 1 \leq \zeta < 2 \\ 0 & \text{otherwise} \end{cases} \quad [1.46]$$

The 2D shape function is then

$$N_i(\zeta, \eta) = \varphi_i(\zeta)\varphi_i(\eta) \quad [1.47]$$

and, given the definition of the local basis variables $\zeta = (\mathbf{x}_p - \mathbf{x}_i) \cdot \mathbf{e}_x/\ell_x$ and $\eta = (\mathbf{x}_p - \mathbf{x}_i) \cdot \mathbf{e}_y/\ell_y$, the derivation of the gradients follows naturally:

$$\boldsymbol{\nabla} N_i(\zeta, \eta) = \begin{pmatrix} \dfrac{\partial \varphi_i(\zeta)}{\partial \zeta} \dfrac{\partial \zeta}{\partial x} \varphi_i(\eta) \\ \varphi_i(\zeta) \dfrac{\partial \varphi_i(\eta)}{\partial \eta} \dfrac{\partial \eta}{\partial y} \end{pmatrix} = \begin{pmatrix} \dfrac{\varphi_i(\eta)}{\ell_x} \dfrac{\partial \varphi_i(\zeta)}{\partial \zeta} \\ \dfrac{\varphi_i(\zeta)}{\ell_y} \dfrac{\partial \varphi_i(\eta)}{\partial \eta} \end{pmatrix} \quad [1.48]$$

– a second solution to mitigate the cell crossing instability is the *generalized interpolation material point* (GIMP) approach proposed by Bardenhagen and Kober [BAR 04]. The idea is to weight the shape function by the part of MP that occupy its support. This weighted shape function formally reads

$$\bar{N}_{ip} = \frac{1}{V_p} \int_{V_p \cap V_e} \chi_p(\mathbf{x}_p) N_{ip} \mathrm{d}V \quad [1.49]$$

where $V_p \cap V_e$ denotes the current support of the characteristic functions χ_p of the MP that defines a "fuzzy" mass distribution instead of a lumped one so that:

$$V_p = \frac{1}{V_p} \int \chi_p(\mathbf{x}_p) \mathrm{d}V \quad [1.50]$$

The weighted shape function plays exactly the same role as the shape function in conventional MPM algorithm. When combined with piecewise linear shape function, however, it lost its C^0 continuity.

There exists thus a similitude between the GIMP and the usage of B-splines. A number of versions are proposed in the literature, e.g. *g*GIMP, *u*GIMP, *cp*GIMP or CPDI, where the differences concern mainly the way the mass distributes around the MP and how it deforms;

– a completely different route to get rid of the cell crossing instability is the *finite element method with lagrangian integration points* [MOR 03]. Here, the strategy is rather to treat the MPs as computational points just like the Gauss points do in the FEM. The MPs are actually replaced by some updated integration points used to approximate the volume integrals in the weak form of the conservation of linear momentum (equation [1.35]). The approximation of a function Ψ involves a weight for each integration points so that:

$$\int_{\Omega} \Psi \mathrm{d}\Omega \simeq \sum w_p \Psi(\mathbf{x}_p) \quad [1.51]$$

The right estimate of the weights w_p affected to the integration points ensures that any quantity is conserved over the domain Ω even if the number of integration points varies in the domain. That way the issues of cell crossing instability is circumvented.

1.3.4.2. *Time discretization*

As usual, the first necessary condition to be satisfied to determine the time step Δt is the Courant–Friedrichs–Lewy condition. In a 2D dynamic case, assuming the maximum velocity is the speed of longitudinal wave in an elastic solid $\sqrt{E/\rho}$, this condition reads:

$$\Delta t < \sqrt{\frac{\rho}{E} \frac{\ell_x \ell_y}{\ell_y + \ell_x}} \quad [1.52]$$

where the density ρ and the Young's modulus E are chosen over the sample so that ρ/E is the smallest.

A second necessary condition is related to the use of DEM-like contact laws to handle the boundary conditions in MPM simulations.

Considering a linear, elastic and non-viscous contact, similarly to DEM simulations, the condition may be:

$$\Delta t < \pi \sqrt{\frac{k_n^{\min}}{m_p^{\max}}} \qquad [1.53]$$

where $k_n^{\min}$ is the smallest normal stiffness of the MP-obstacle contact and $m_p^{\max}$ is the heaviest MP mass. Obviously, the employed time increment has to be a portion (typically a tenth) of the smallest time increments given by equations [1.52] and [1.53].

1.3.5. *Constitutive model*

A constitutive model aims at defining a relation between a stress increment and a strain increment. The evaluation of the strain tensor increment $\Delta\varepsilon$ is thus required as an input for each MP. Note that the italic subscript ($_p$) related to the MPs is not used in this section; this is to avoid some possible confusion with the plastic term of strain that will rather be superscripted with the non-italic letter (^{P}) in what follows. The strain tensor increment can be obtained from the strain rate tensor ($\Delta\varepsilon = \dot{\varepsilon}\Delta t$), which is the anti-symmetric part of the velocity gradient tensor, and is equated as follows:

$$\dot{\varepsilon}(t) = \frac{1}{2}\sum_i \Big(\mathbf{v}_i(t) \otimes \boldsymbol{\nabla} N_{ip}(t) + \boldsymbol{\nabla} N_{ip}(t) \otimes \mathbf{v}_i(t)\Big) \qquad [1.54]$$

Considering Hook's elasticity, the loading increment is

$$\Delta\boldsymbol{\sigma} = \frac{E}{1+\nu}\left(\Delta\varepsilon + \frac{\nu}{1-2\nu}\,\mathrm{tr}(\Delta\varepsilon)\,\mathbf{I}\right) \qquad [1.55]$$

where E is the Young modulus, ν is the Poisson ratio and $\mathbf{I}$ is the identity matrix. For convenience, the engineering representation of the stress and strain is introduced:

$$\tilde{\boldsymbol{\sigma}} = {}^{T}\!(\sigma_{xx}\ \sigma_{yy}\ \sigma_{zz}\ \sigma_{xy}\ \sigma_{xz}\ \sigma_{yz}) \qquad [1.56]$$

$$\tilde{\varepsilon} = {}^{T}\!(\varepsilon_{xx}\ \varepsilon_{yy}\ \varepsilon_{zz}\ 2\varepsilon_{xy}\ 2\varepsilon_{xz}\ 2\varepsilon_{yz}) \qquad [1.57]$$

With this notation, Hook's relation expresses

$$\Delta\tilde{\sigma} = \mathbf{D}^{\mathrm{e}} \cdot \Delta\tilde{\varepsilon}^{\mathrm{e}} \qquad [1.58]$$

where $\mathbf{D}^{\mathrm{e}}$ is the classical elastic operator.

For landslides or rock avalanches, it is essential to model plasticity. We present here the two-dimensional case of Mohr–Coulomb plasticity that involves an open bilinear yield surface with an apex for purely tensile stresses, a non-associated flow rule, constant rate of volume change and no hardening neither softening. In this peculiar case, the yield surface f is expressed as a function of the stress components:

$$f(\tilde{\sigma}) = \sqrt{4\sigma_{xy}^2 + (\sigma_{xx} - \sigma_{yy})^2} + (\sigma_{xx} + \sigma_{yy}) \sin\varphi$$
$$-2c \cos\varphi \qquad [1.59]$$

where φ is the angle of internal friction and c is the cohesion. The plastic potential g reads

$$g(\tilde{\sigma}) = \sqrt{4\sigma_{xy}^2 + (\sigma_{xx} - \sigma_{yy})^2} + (\sigma_{xx} + \sigma_{yy}) \sin\psi$$
$$-2c \cos\psi \qquad [1.60]$$

where ψ is the dilatancy angle (with $\psi \leq \varphi$). Figure 1.12 gives an illustration of this yield surface in the space τ_{m}–σ_{m} where the isocurve $f(\tilde{\sigma}) = 0$ is

$$\tau_{\mathrm{m}} = \sigma_{\mathrm{m}} \tan\varphi + c \qquad [1.61]$$

with

$$\tau_{\mathrm{m}} = \frac{\sqrt{4\sigma_{xy}^2 + (\sigma_{xx} - \sigma_{yy})^2}}{2\cos\varphi} \qquad [1.62]$$

$$\sigma_{\mathrm{m}} = -\frac{\sigma_{xx} + \sigma_{yy}}{2} \qquad [1.63]$$

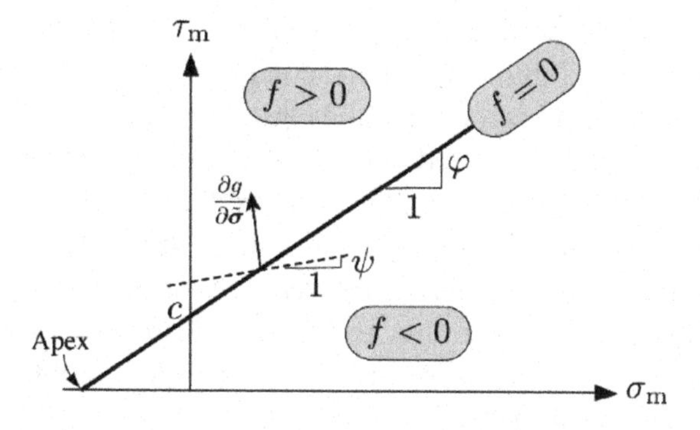

Figure 1.12. *Yield surface*

As usual in computational plasticity, the following fundamental rules need to be satisfied:

$$\begin{cases} \tilde{\varepsilon} = \tilde{\varepsilon}^{\mathrm{e}} + \tilde{\varepsilon}^{\mathrm{p}} \\ \Delta\tilde{\varepsilon}^{\mathrm{p}} = \Delta\lambda \dfrac{\partial g}{\partial \tilde{\sigma}} \\ f(\tilde{\sigma}) \leq 0 \end{cases} \qquad [1.64]$$

and this is accomplished by means of a return algorithm involving a Newton–Raphson iterative procedure, which is summarized in the pseudo-code below:

1: Initial stress state: $\tilde{\boldsymbol{\sigma}}_{(k=0)}$
2: **repeat**
3: Compute the plastic multiplier:
$$\Delta\lambda = \frac{f(\tilde{\boldsymbol{\sigma}}_k)}{\left(\dfrac{\partial f}{\partial \tilde{\boldsymbol{\sigma}}_k}\right)^T \cdot \mathbf{D}^{\mathrm{e}} \cdot \dfrac{\partial g}{\partial \tilde{\boldsymbol{\sigma}}_k}}$$
4: Compute the plastic predictor:
$$\Delta\tilde{\boldsymbol{\sigma}}^{\mathrm{p}} = -\Delta\lambda\, \mathbf{D}^{\mathrm{e}} \cdot \frac{\partial g}{\partial \tilde{\boldsymbol{\sigma}}_k}$$
5: Update the stress state:
$$\tilde{\boldsymbol{\sigma}}_{k+1} = \tilde{\boldsymbol{\sigma}}_k + \Delta\tilde{\boldsymbol{\sigma}}^{\mathrm{p}}$$
6: **until** $(f(\tilde{\boldsymbol{\sigma}}) \leq \epsilon$ or $k > k_{\max})$

It must be noted that the return algorithm explained above is notably used to fulfill the yield condition, and since both the Mohr–Coulomb yield surface and the plastic potential are linear, only a single step is actually required in the Newton–Raphson loop. However, the return toward the apex of the yield surface still requires a certain number of iterations. This is precisely where the difficulty lies when dealing with very large straining and very low stresses at the surface of the mass flow. Special care has to be taken in order to return to the apex correctly. The solutions to overcome this problem are not developed here for the sake of brevity.

It is important to understand that all the existing developments dedicated to FEM are still usable in the framework of the MPM. In particular, more sophisticated constitutive model implying hardening, softening, limited volume changes and so on can be implemented with no additional difficulties.

2

DEM Applied to Laboratory Experiments

This chapter aims to test the ability of the DEM to simulate well-controlled experiments that involve particle flows. Two experiments were chosen from the literature for this purpose: channelized avalanches of plastic pellets or glass beads [SAV 91, HUT 91, HUT 95], which can be assumed to be two-dimensional (2D) experiments, and a three-dimensional (3D) experiment carried out at the EPFL (École polytechnique fédérale de Lausanne) [MAN 08] involving small baked-clay bricks. The main difficulties met when comparing some experimental and numerical results relate to the definition of adequate parameters or in the ability of the numerical model to account for all specific experimental characteristics (boundary conditions, right number and realistic shapes of particles for example). The experiments of Savage and Hutter [SAV 91], Hutter and Koch [HUT 91] and Hutter *et al.* [HUT 95] were chosen because of the fact that the authors have made several attempts to measure the main physical parameters such as normal dissipation or friction coefficients, which will necessarily be used as input parameters in the modeling, and because the experiments can be considered as 2D problems. Nevertheless, due to the large number of particles and the complex shape of the elements used in the experiments (for plastic pellets in particular), it was not possible to model all details in the performed experiments. In this case, only the global macroscopic behavior of the granular flow is of interest; analyzing it from a microscopic viewpoint would be misleading. In contrast, the experiments performed at the EPFL [MAN 09] were chosen because of their simple geometric configuration (low number of particles, simple boundary conditions, regular sizes and well-controlled shapes of the particles). Also, the materials used for the experiments (bricks and plastic coating for the propagation area) have been made available to us because of a collaboration established as part of the European project Alcotra MASSA (Medium and Small Size Rock Fall Hazard Assessment, 2010–2013) in order to assess the dissipation parameters via elementary rebound tests with single bricks.

The work presented in this chapter is the result of contributions from Julien BANTON, Dominique DAUDON and Guilhem MOLLON.

2.1. Description of the experiments

2.1.1. *2D releases within a channel*

These experiments carried out by [SAV 91] and [HUT 95] were performed within a 0.1 m wide channel comprising (1) a rectilinear plan of 1.4 m in length with an inclination angle α ranging from 50° to 60°, (2) a curved section with a curvature radius $R_c = 0.246$ m and (3) a horizontal plan (Figure 2.1).

Two kinds of particles were used: lens-like shaped plastic particles called Vestolen (950 kg·m^{-3}; Figure 2.1(a)), and nearly spherical glass beads (2,860 kg·m^{-3}; Figure 2.1(b)). The plastic pellets were 4 mm in diameter and 2.5 mm in height, and the glass beads were about 3 mm in diameter. The masses M of material were poured into a container positioned at the top of the channel with an apparent density of 540 kg·m^{-3} for the plastic particles and 1,730 kg·m^{-3} for the glass beads. The whole mass was then released by rapidly opening a gate that rotates around a horizontal axis, and which is sited at the bottom of the container (Figure 2.1(c)). The side walls of the channel have low friction with the particles in order to limit their effect on the flow and to enable the assumption of a 2D flow. During the flow, the granular mass was photographed seven times per second for assessing the kinematic of the granular flow together with the shape and position of the final deposit. Two different coating materials were used for the bed linings: a drawing paper and a sand paper. Among all the experiments performed [SAV 91, HUT 95], only the three physical experiments, presented in Table 2.1, were selected for comparison with the DEM.

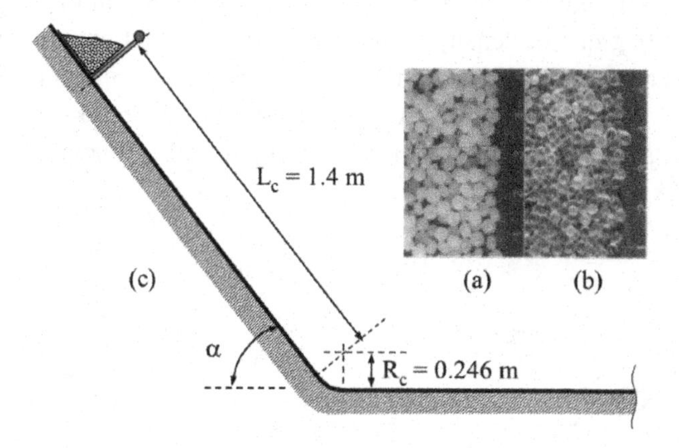

Figure 2.1. *Materials and geometry of the experiment performed by Hutter et al. [HUT 95]: a) plastic pellets, b) glass beads and c) experimental channel*

Experiment	α (°)	Coating	Material	Particle number	Particle weight (g)
#A1	50	Drawing paper	Vestolen	50,000	1,500
#A2	50	Sand paper	Vestolen	67,000	2,000
#A3	60	Drawing paper	Glass	100,000	4,000

Table 2.1. *Characteristics of the experiments performed by Hutter [HUT 91] and Hutter et al. [HUT 95]*

2.1.2. *3D releases on a two-side slope*

These experiments [MAN 09], performed at the EPFL, consisted of the release and the flow of an assembly of small bricks on a biplan (Figure 2.2). The apparatus is composed of two rectangular boards (3 × 4 m) coated with a smooth plastic material (called forex) and linked together by a sharp hinge. The first board is inclined at 45°, while the second one is horizontal. A rectangular container (0.2 m height, 0.4 m width and 0.6 m long) positioned at the desired height on the inclined plan was filled with 40 L of bricks (Figure 2.2). The lower part of the container has a moving gate that allows the release of the brick assembly. Two experiments were carried out using random or ordered assemblies of small clay bricks $31 \times 15 \times 8$ mm in dimension and $1,700$ kg·m^{-3} in density (Figure 2.3). For the first experiment, the bricks were poured randomly into the container at an apparent density of 1,000 kg·m^{-3} while for the second experiment, the bricks were piled meticulously to reach the maximum volume of 40 L (apparent density of 1,600 kg·m^{-3}).

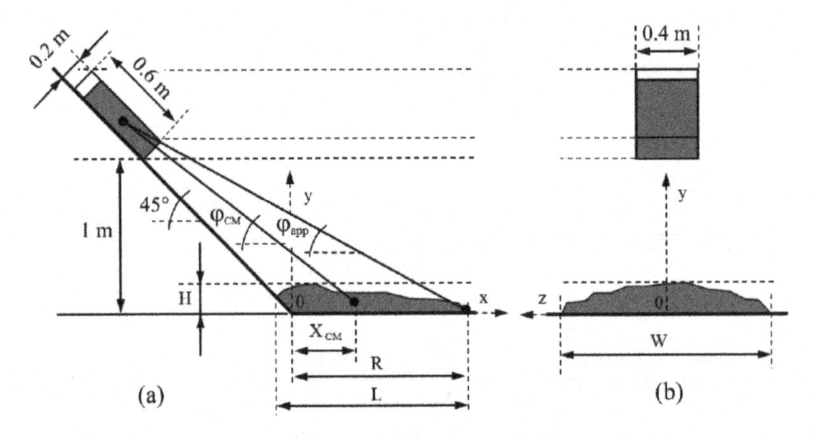

Figure 2.2. *Geometries of experimental apparatus and brick deposit: a) lateral view and b) front view [MAN 09]*

A high-speed camera was used during the experiences to evaluate the position and the velocity of the mass front. The morphology and the dimensions of the material deposit (i.e. dimensions, position and shape) were determined by an optical technique (a fringe projection technique). The main parameters defining the dimensions and the position of the deposit are the length L, the runout distance R, the width W, the thickness H, the position of the mass center X_{CM} and the angles of propagation (travel angle φ_{CM} related to the mass center, and Fahrböschung angle φ_{app} related to the front of the deposit).

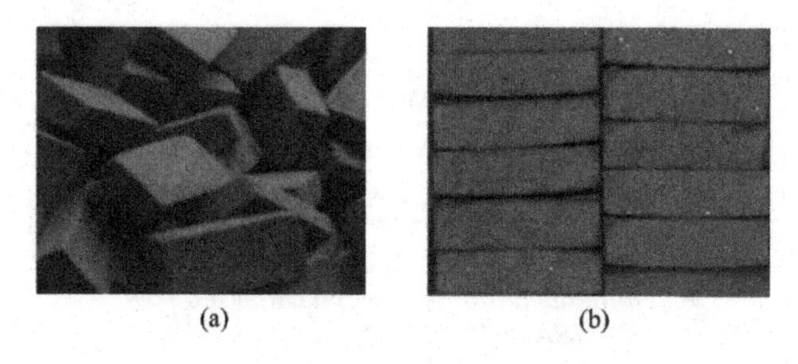

Figure 2.3. *Randomly a) or piled b) assemblies of bricks used by Manzella and Labiouse [MAN 09]*

2.2. Definition and assessment of the contact parameters

2.2.1. *2D releases*

The elastic properties of the materials used for the experiments #A1 to #A3 (Vestolen particles, glass beads, drawing paper and sand paper) were determined on the basis of specific laboratory tests carried out by Savage and Hutter [SAV 91], Hutter and Koch [HUT 91] and Hutter *et al.* [HUT 95]. The different stiffnesses were set accordingly. The Coulomb friction angles acting in-between the granular materials (Vestolen particles and glass beads) were correlated [HUT 95] to the static angle of repose (φ_i) of a wedge-type pile of material deposited on a horizontal frictional plan covered with drawing paper or sand paper – according to the case. The angles of repose were accurate to $\pm 4°$ owing to the effect of scoring and measurement inaccuracy. Bed friction angles (φ_{bed}) between granular materials and pieces of drawing paper or sandpaper were determined using an inclined plan that tilts gradually until the granular material was destabilized. The corresponding tilt angles were assumed to be the bed friction angles with a margin of uncertainty of about $2°$. The normal restitution coefficients e_n were obtained using tests of fall and rebound of a particle on a horizontal plan covered with pieces of drawing paper or sandpaper. The coefficient is defined by $e_n = \sqrt{h_1/h_0}$ where h_0 is the height of release and h_1 is the height of the following rebound. The physical properties of the Vestolen particles, glass beads, drawing paper and sand paper are summarized in Table 2.2.

Experiment	Coating	Material	φ_{bed} (°)	φ_i (°)	e_n (-)
#A1	Drawing paper	Vestolen	19.0	29.0	0.61
#A2	Sandpaper	Vestolen	28.5	33.5	0.54
#A3	Drawing paper	Glass	26.0	28.0	0.48

Table 2.2. *Physical properties of the material used for the 2D experiments performed with a canalized flow [SAV 91, HUT 91, HUT 95]*

2.2.2. *3D releases*

Concerning the 3D releases of small bricks on a two-sided slope, the physical parameters of the contact law were assessed by analyzing the fall and rebound of a single brick on a horizontal plan. Four brick–support collisions and two brick–brick collisions were analyzed. The contact parameters were identified for each type of contact. For brick–brick collisions, the basal support was made up of several bricks orderly stuck on the support, while for the brick–support collisions, the smooth plastic material (forex) was used. The materials of the experiments (bricks and coating material) were the same as the original ones used by Manzella and Labiouse [MAN 09]. The fall, the collision, and the bouncing were shot at 1,000 frames per second using two high-speed cameras positioned along two orthogonal directions. The optimization uses a *brute-force* approach, which involves the following steps: (1) experimental identification of actual trajectory of the brick, (2) numerical computation of the brick trajectories by varying the mechanical parameters in a certain range, and (3) identification of the best set of parameters.

2.2.2.1. *Experimental identification of brick trajectory*

Experimental trajectory of one brick can be obtained by following, during its fall and its rebound, the motion of four characteristic points located at different visible corners of the brick (referred to as A_1, A_2, A_3 and A_4 on Figure 2.4). The points A_1 and A_2 are visible from both cameras, while the points A_3 and A_4 only appear on cameras 2 and 1, respectively.

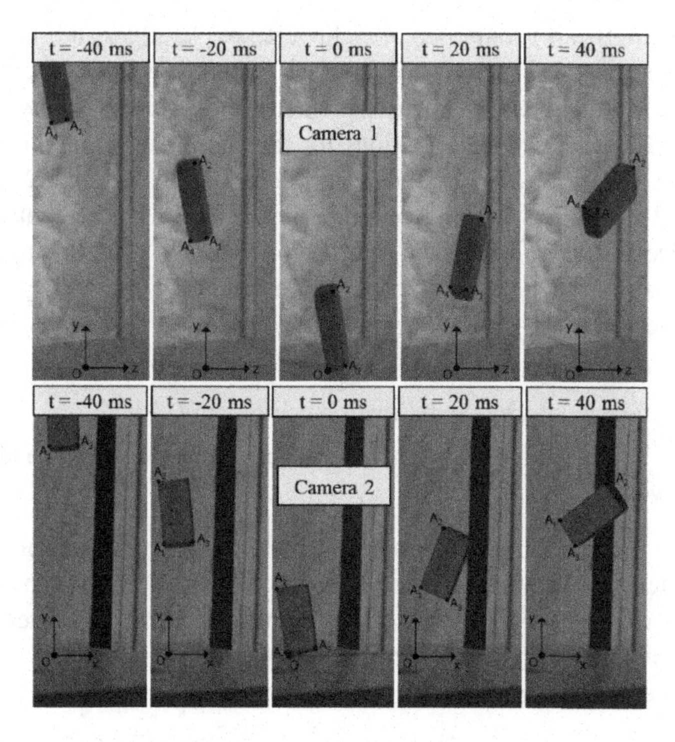

Figure 2.4. *A view from two orthogonal cameras of the brick during its fall and bouncing [RIC 12]*

For each of the 200 pairs of images shot by the cameras, the positions of the points (assessed with a subpixel accuracy) on the images are estimated using a digital image correlation technique. This operation provides the trajectories of A_1, A_2 and A_4 (respectively, A_1, A_2 and A_3) projected on camera 1 (respectively, on camera 2), with a time resolution of 1 ms. After synchronization of the 2D trajectories of each corner (the collision time is set as the time origin by extrapolation) and scaling (size of a pixel for each camera), it was possible to reconstruct the actual 3D trajectory of each brick.

2.2.2.2. *Numerical computation of the brick trajectory*

The numerical simulations consisted of simulating the fall and bouncing of a single brick on a horizontal plan. The numerical brick ($31 \times 15 \times 8$ mm) is modeled using a spheropolyhedron (the Minkowski radius being 1 mm) composed of eight vertexes, 12 edges and six faces. The experimental trajectories of the brick being known, the parameters of the fall (initial position, rotation, velocity and angular velocity of the brick) are used as initial conditions for the simulation. The numerical tool and the contact law used are those described in section 1.2. Since neither the support nor the bricks are soft, it was not required to introduce rolling resistance so that the only means of dissipation are collisions and friction. The contact parameters for the two types of interaction are:

– the coefficient of normal energy restitution (e_n^2);

– the tangential friction dissipation coefficient (μ);

– the normal and tangential contact stiffnesses (k_n and k_t, respectively).

For a given set of the four contact parameters, the model provides a trajectory of the brick after the collision, which may be compared to the experimental one. When using the optimal set of parameters, the numerical trajectory fits closely with the experimental trajectory for the two cameras (Figure 2.5).

2.2.2.3. *Optimization procedure to identify the parameters*

The optimization procedure consists of varying the four numerical parameters (e_n^2, μ, k_n and k_s) and in calculating, for each set of parameters, an error function defined as a sum over time of the distances between the positions of the corners of the brick obtained experimentally and numerically. For each couple of parameters of the contact law, the isovalue curves of the error function make it possible to identify, for each experimental test, the most relevant values of the parameters needed to restore the experimental trajectory after the collision. Assuming that the collision velocity and its localization (corners, edges or faces) have a minor influence on the values of the contact parameters, it is possible to obtain, considering all the tests performed together, an optimal set of parameters. As an example, we present in Figure 2.6 the isovalue curves of the error function referring to the parameters e_n^2 and μ for a collision implying a brick and the forex material. In the blue central part of Figure 2.6, the parameter values that best minimize the error function are $e_n^2 = 0.53$ and

$\mu = 0.46$. Taking account of all the tests performed, it was possible to define the optimal sets of parameters for the brick–support and brick–brick interactions; these are given in Table 2.3.

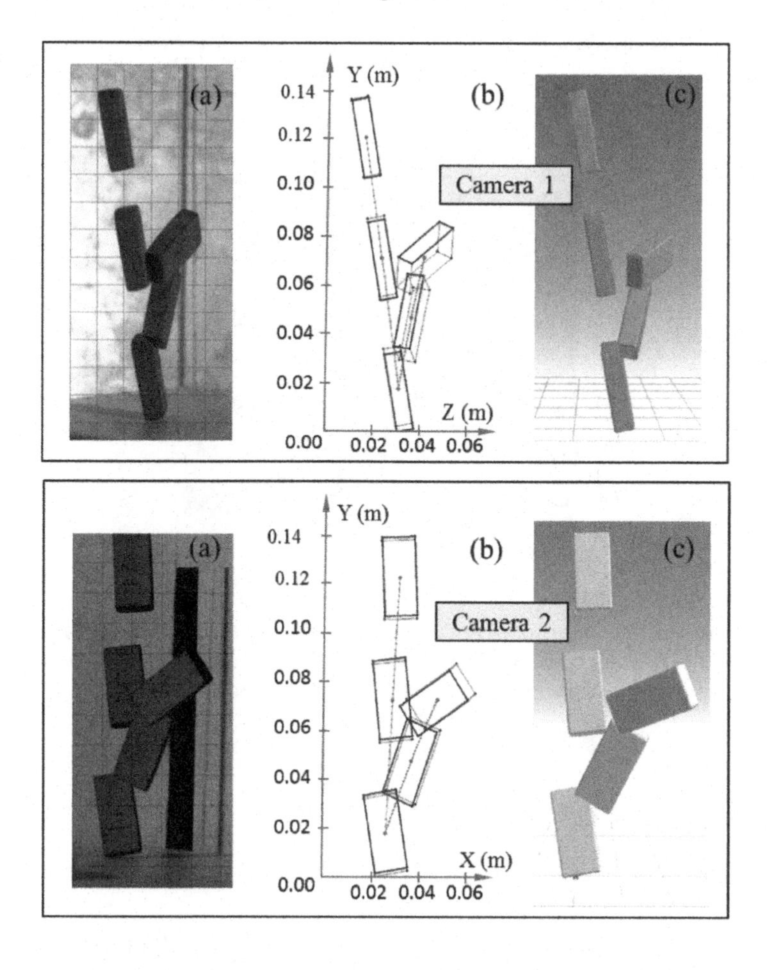

Figure 2.5. *Images shot by the two cameras every 20 ms a), experimental trajectories identified from image analysis b), numerical trajectories obtained by optimizing the contact parameters c) [RIC 12]*

Interaction type	e_n^2	μ	k_n (N/m)	k_s/k_n
Brick–support	0.53	0.46	10^5	0.42
Brick–brick	0.13	0.86	10^5	0.27

Table 2.3. *Optimal sets of parameters for brick/support and brick/brick interactions*

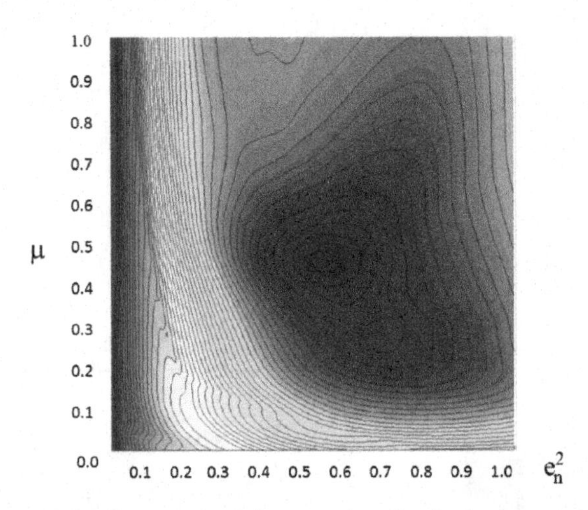

Figure 2.6. *Isovalue curves of the error function in the parameter space e_n^2–μ for the collision of a brick with the forex support. For a color version of this figure, see www.iste.co.uk/richefeu/gravity.zip*

2.3. Simulation versus experiment results

2.3.1. *2D releases*

Considering the canalized channel experiment as a 2D problem is a strong assumption, which was necessary in order to minimize the number of particles and the numerical simulation duration. Furthermore, this assumption imposes a simplification of the geometry of the particles that are described in 2D. Due to the lack of information about the normal contact dissipation between two granular particles (Vestolen pellets and glass beads), only the friction dissipative mechanism was considered within the granular mass. In contrast, a dissipative contact law similar to the one described previously (section 1.2.6) was used at the base of the flow in order to take into account the frictional and normal energy dissipation between particles and bed linings. The discrete-element software used for this study is the 2D code PFC^{2D} [ITA 96]. The granular material was modeled using a collection of elementary discs of diameter D that were assembled together to form clusters [BAN 09]. Discs were used to represent the glass beads while clusters of two overlapping discs of same diameter (from which the centers are separated by a distance of 0.6 D) were used to roughly mimic the shape of the plastic pellets. An amount of 1839, 1740 and 4042 particles were used to simulate experiments #A1 to #A3, respectively. Considering the channel width (0.1 m) and the nominal diameters of plastic pellets (4 mm) and glass beads (3 mm), the number of 2D particles approximately coincides with the number of the 3D physical experiment particles ($50,000$ divided by 25 for experiment #A1, $67,000$ divided by 25 for experiment #A2 and $100,000$ divided by 33 for experiment #A3). The numerical particles are generated randomly at fixed porosity in the container. Gravity is applied and is followed by a first stage of stabilization of the granular mass under its own weight. The gate sited at the bottom of the container was suppressed allowing the propagation of the granular material along the slope and its stabilization on the horizontal plane. The displacement boundary conditions at the base of the flow were imposed by rigid frictional walls. The contact parameters were deducted from the experimental characteristics given in Table 2.2. The bed friction parameter μ was defined as a function of the bed friction angle ($\mu = \tan \varphi_{\text{bed}}$). The Coulomb friction parameter μ between two granular particles was deduced indirectly from additional numerical simulations that were performed to reproduce the angle of repose of a pile of numerical granular material placed on a frictional horizontal plane. Table 2.4 summarizes the contact parameters used in the three simulations.

The comparison between the experimental and numerical results [BAN 09] is given in Figures 2.7–2.9 in terms of the kinematics of the flow for the three experiments performed. As it can be seen on these figures, despite the fact that all the physical phenomena were not taken into account in the numerical model, and considering all uncertainties on both data and input parameters, a remarkable agreement was obtained for the three experiments for the deposit shapes at rest. Additionally, the computed mass propagation at different times is quite consistent with the experimental snapshots regardless of the shape of the particles implied in the granular flow. Indeed, as we will see in the following chapters, the main parameters that influence the granular flow and the shape and position of the deposit are the particle shape and the dissipative coefficients at the basal surface. Defining these parameters well during the experiment campaign explains the agreement between the numerical and experimental results. More precisely, Figure 2.7 compares the experimental and modeled snapshots at 11 propagation times of experiment #A1 [BAN 09]. Although the initial profile shapes are slightly different, the mass propagates down the slope in the same way as the time progresses. After stabilization of the numerical plastic pellets, it can be seen that the deposits exhibit the same length and asymmetric shape.

Interaction type	e_n^2	μ	k_n (N/m)	k_s/k_n
Vestolen–drawing paper	0.37	0.34	1.5×10^8	1
Vestolen–sandpaper	0.29	0.54	1.5×10^8	1
Glass–drawing paper	0.23	0.49	1.5×10^8	1
Vestolen–Vestolen	1.00	0.54	1.5×10^8	1
Glass–lass	1.00	0.49	1.5×10^8	1

Table 2.4. *Numerical contact parameters for interactions between Vestolen, glass, drawing paper and sandpaper [BAN 09]*

Figure 2.8 compares the experimental and simulated snapshots at four different times of experiment #A2 [BAN 09]. Compared with experiment #A1, the runout distance and the deposit length of the numerical plastic pellets are shorter, and the deposit shape is strongly asymmetrical, because of the higher bed friction angle (28.5° instead of 19°). The only discrepancy is the slight delay observed between the experimental and simulated rear ends at the last stage of the experiment.

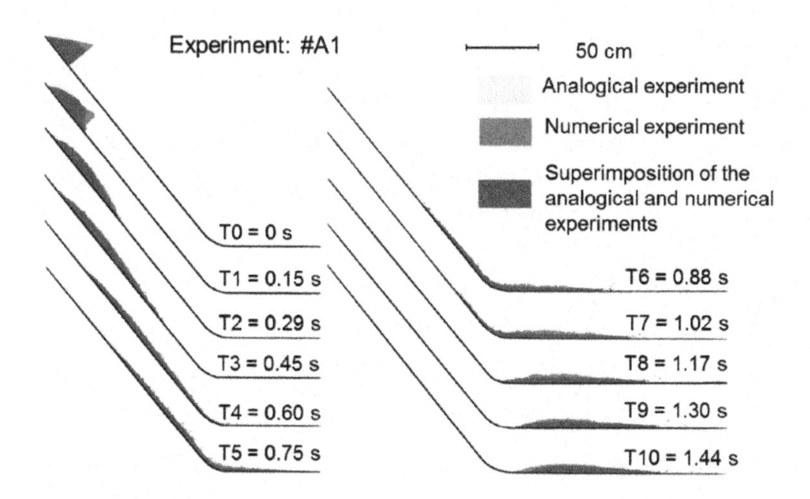

Figure 2.7. *Comparison between snapshots of experiment #A1 (yellow) [SAV 91] and DEM simulations (orange) [BAN 09] at 11 travel times. Common parts are displayed in red. For a color version of this figure, see www.iste.co.uk/richefeu/gravity.zip*

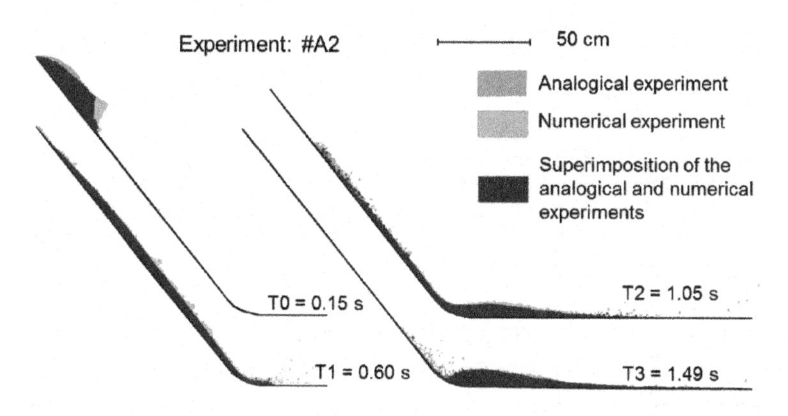

Figure 2.8. *Comparison between snapshots of experiment #A2 (gray) [SAV 91] and DEM simulations (light blue) [BAN 09] at four travel times. Common parts are displayed in dark blue. For a color version of this figure, see www.iste.co.uk/richefeu/gravity.zip*

Experimental snapshots of experiment #A3 are given in Figure 2.9 [BAN 09]. As expected for this experiment, the runout distance and the spreading are much larger than in the two previous experiments because of the circular particle shape, the relatively low bed friction angle and a greater inclination angle between the slope and the horizontal plane. Also, the obtained deposit shape is more symmetrical.

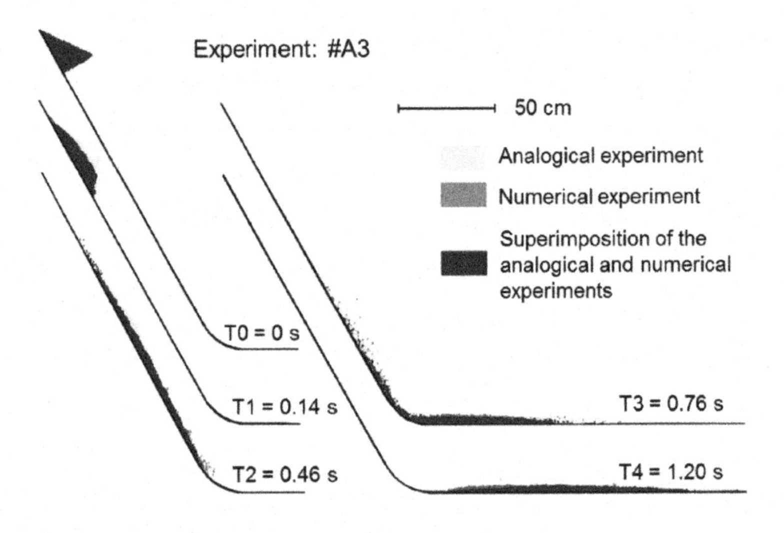

Figure 2.9. *Comparison between snapshots of experiment #A3 (green) [SAV 91] and DEM simulations (dark green) [BAN 09] at five travel times. Common parts are in brown. For a color version of this figure, see www.iste.co.uk/richefeu/gravity.zip*

Considering the strong agreement between the experimental and numerical results obtained for the three experiments (using different particle shapes or various bed friction materials), these results show the ability of the DEM model to accurately reproduce the kinematics of granular flows after meticulously defining the numerical parameters.

2.3.2. *3D releases*

2.3.2.1. *Randomly poured bricks*

These numerical simulations refer to the experiment from [MAN 09] performed with 40 L of randomly poured clay bricks. The geometry of the experimental apparatus and the size and position of the container are those presented previously in section 2.1.2 in Figure 2.2. The values of the contact parameters are those determined after the calibration process implying the fall and bounce of single bricks (Table 2.3). The 6,300 bricks were laid out on the container at an apparent density of $1,000$ kg·m^{-3}, each brick being oriented in a random direction. The propagation of the bricks along the inclined plan started when the lower face of the container was suppressed. The kinematics of the flow and a view of the numerical deposit are shown in Figure 2.10 while a comparison between the experimental and numerical results is given in Figure 2.11 and Table 2.5 in terms of (1) contour of the deposit (in plan and in elevation), (2) mass-front horizontal velocity and (3) main characteristics of the numerical and experimental deposits. Figure 2.10 shows that the granular flow progressively spread laterally during the flow along the inclined plan and finally stops on the horizontal plan. The detailed view of the final deposit (Figure 2.10) shows that some bricks are ejected from the principal mass and that the thickness of the deposit is higher in the central part.

As shown in Table 2.5, the comparisons between the numerical and experimental deposit positions are rather satisfactory. In particular, the position of the center of mass of the deposit (X_{CM}), the travel angle (φ_{CM}) and the Fahrböschung angle (φ_{app}) related to the front of the granular deposit are very well replicated by the numerical model. Moreover, Figure 2.11(a) shows a very satisfying correspondence between the numerical and experimental deposits projected in the horizontal plan (top-view). However, some differences can be observed on the elevation views (side-view) of the brick deposit (Figure 2.11(b)) that may be related to a concentration of bricks in the central part of the numerical deposit (Figure 2.10). The front velocity versus the position of the front (Figure 2.10(c)) shows a rather good match of the experimental and numerical results; in this plot, we need to take care that the experimental velocities have been obtained with a lower accuracy than the numerical ones. From both experimental and numerical results, a net deceleration can be observed when the mass front reaches the horizontal plane (mass front position X in the range of 0–0.2 m), followed by a gentler decrease in the velocity, corresponding to the accumulation of the material on the plane (X ranging from 0.2 to 0.6 m). Just before the mass stops, a second steep deceleration (X ranging from 0.6 to 0.8 m). The rather good accordance of the numerical model with the experiment may be attributed to the fact that the collective behavior of the bricks is less sensitive to the micromechanical contact parameters. On the contrary, the trajectory of a single brick is largely governed by the local contact conditions at each bounce, i.e. geometry of the collision (corner, edge or face) or micromechanical parameters.

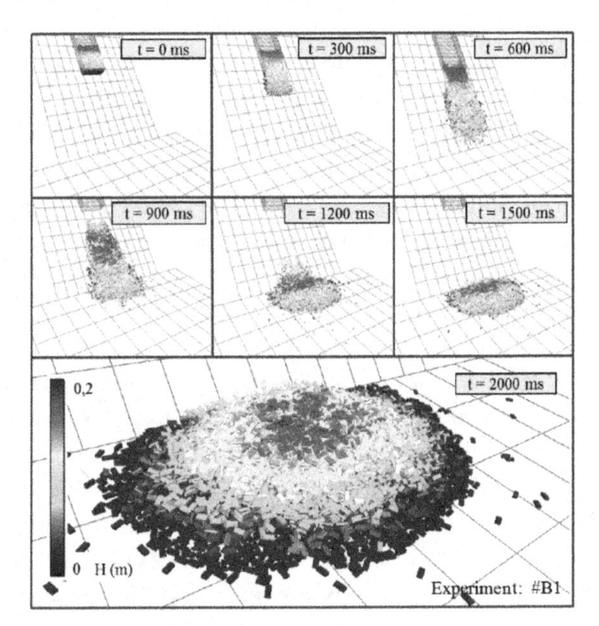

Figure 2.10. *Kinematics of the granular flow (top) and detailed view of the final deposit (bottom) in the case of randomly poured brick release (experiment #B1). For a color version of this figure, see www.iste.co.uk/richefeu/gravity.zip*

	φ_{CM} (°)	φ_{app} (°)	L (m)	R (m)	W (m)	H (m)	X_{CM} (m)
Experiment #B1	40	32	0.93	0.84	1.40	0.075	0.37
Simulation #B1	40.1	32.2	0.88	0.82	1.38	0.120	0.33

Table 2.5. *Comparison between experimental and numerical characteristics of the deposit in the case of randomly poured brick experiment (experiment #B1)*

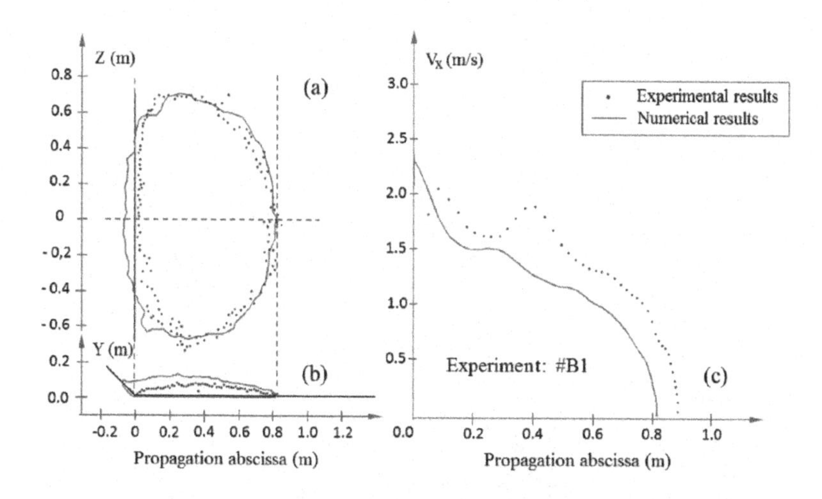

Figure 2.11. *Contours of the final deposit: top-view a) and side-view b). Velocity of the mass front as a function of the position of the front mass c) in the case of randomly poured brick release (experiment #B1)*

2.3.2.2. *Orderly piled bricks*

These numerical simulations refer to the experiment from [MAN 09] involving 40 L of clay bricks that have been regularly and meticulously aligned in the container. Because of the irregularities in the shapes of the bricks as we can observe in Figure 2.3(b), the numerical shape of bricks was adapted by introducing a bevel of 2° on the three face types. About 10, 000 bricks were piled regularly but with their bevel faces randomly positioned, so that it does not trigger a too strong asymmetry in the flow. The kinematics of the flow and a view of the numerical final deposit are shown in Figure 2.12. Compared to previous results obtained with randomly poured bricks (Figure 2.10), orderly piled bricks lead to a higher, longer and narrower numerical deposit; also, the lateral spreading of the granular mass during the flow is less pronounced. This is due to the sliding movements of the bricks within the mass and at the base of the flow: the rotations are limited, and thus scarcer, thanks to the initial ordered arrangement of the bricks.

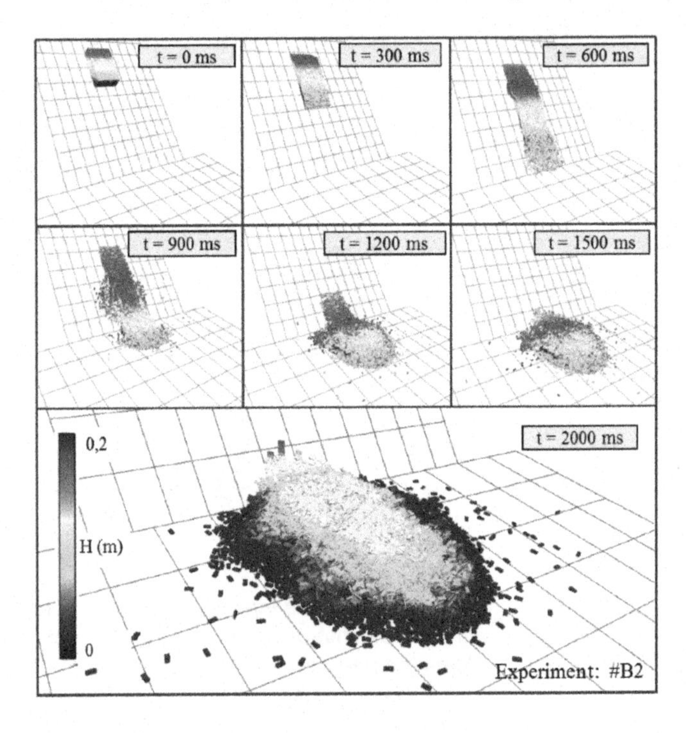

Figure 2.12. *Kinematics of the granular flow (top) and detailed view of the final deposit (bottom) in the case of orderly piled brick release (experiment #B2). For a color version of this figure, see www.iste.co.uk/richefeu/gravity.zip*

The morphology of the deposit and the front velocity are shown in Figure 2.13 for both the experiment and the simulation #B2. Again, the simulation matches well with the experimental observations (shape of the deposit and kinematic of the stop). The net deceleration arising when the mass front reaches the horizontal plane is not apparent in the experimental data (although it has been spotted with randomly poured bricks), while it is clearly perceived with the numerical data. It is also interesting to note that the order of magnitude of the front velocities are rather larger for orderly piled bricks than for randomly poured bricks. This is once again due to the sliding mechanisms that occur in place of frictional and collisional interactions between the bricks.

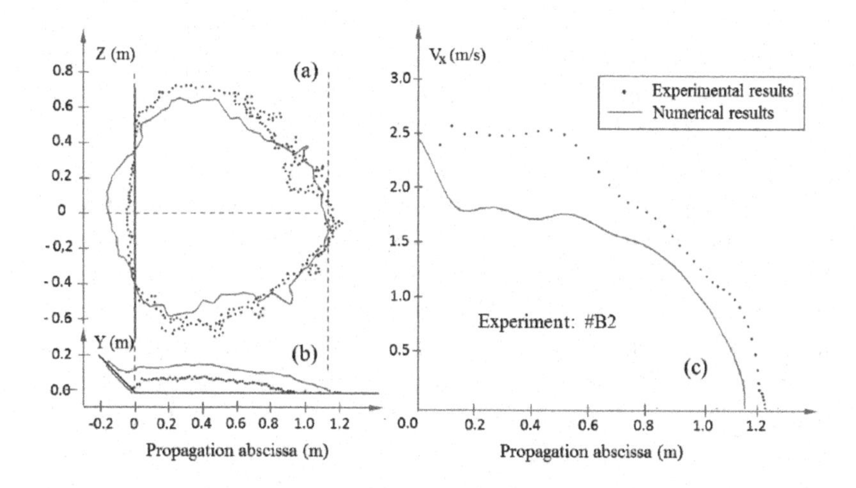

Figure 2.13. *Contours of the final deposit: a) top-view, b) side-view. c) Velocity of the mass front versus the position of the front mass in the case of orderly piled bricks*

	φ_{CM} (°)	φ_{app} (°)	L (m)	R (m)	W (m)	H (m)	X_{CM} (m)
Experiment #B2	39	28	1.23	1.15	1.39	0.089	0.39
Simulation #B2	38.1	29	1.29	1.12	1.21	0.163	0.42

Table 2.6. *Comparison between experimental and numerical characteristics of the deposit in the case of orderly piled brick release (experiment #B2)*

2.4. Further clues handled by numerical results

One asset of the numerical model is its ability to give complementary insights that cannot be measured or determined experimentally. This is the case, for example, for the modes of energy dissipation and some specific aspect of the granular flow that are addressed in the following sections.

2.4.1. *Dissipation modes*

The analysis of the modes of energy dissipation during the propagation is a relevant investigation method for understanding the physical mechanisms involved. Knowing the contact forces and the relative displacements between the particles themselves or between the particles and the support, it is possible to compute the work of contact forces and to deduce the amount of energy dissipated along the time or at each point of the propagation zone, either within the granular mass or at the base of the flow. The energy dissipation may be decomposed in four categories: brick–support frictional dissipation, brick–support normal dissipation, brick–brick frictional dissipation and brick–brick normal dissipation. Taking account of the rates of dissipated energies in the time course and along space, it is possible to obtain relevant information about the loci and intensity of various dissipation modes.

As an example, the amounts of dissipated energy on each section of the x-axis of the propagation path (each slice having a width of 0.05 m) are represented in Figure 2.14 for the experiment #B1 involving randomly poured bricks. Apart from the transition zone between the two planes, the energy mostly dissipates by basal friction (nearly 90% of the total energy dissipated in these areas). In contrast, most of the energy dissipated in-between the bricks (either by friction or by collisions) arises at the breaking of slope and may be related to the chaotic motion of the bricks induced by the brutal change in the orientation of the flow. The same trend is observed at the very beginning, just after the release of the granular mass, where the front bricks fall down the slope.

The change over time of the partition of the different kinds of energies (potential energy, kinetic energy, dissipated energy) of the granular system are plotted in Figure 2.15. Before the release of the bricks at $t = 0$, the static system only stores a given potential energy (light purple in Figure 2.15). When the flow develops along the inclined plan (from $t = 0$ to $t = 0.64$ s), this amount of potential energy declines by transforming into (1) the kinetic energy of the flowing particles (different levels of blue in Figure 2.15) and (2) the dissipated energies resulting from the collisions and the friction at the base or within the granular mass (hot colors ranging from yellow to red in Figure 2.15). Soon after the inlet of the avalanche on the horizontal portion, the kinetic energy reaches a peak and decreases until the motion ceases, at $t \simeq 1.4$ s. In the total event, most of the energy was dissipated by friction between the support and the bricks (red, 66.2%) and friction between bricks (dark orange, 24.2%). The energy dissipated by normal damping is much less significant. This is a direct consequence of the sliding motion of the brick mass on the inclined slope and of the change of the flow nature, which was abruptly disturbed at the transition zone between the two plans.

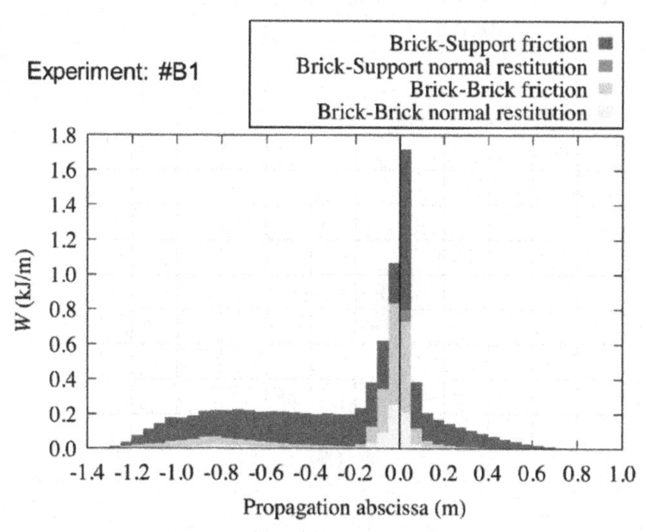

Figure 2.14. *Cumulated energies by unit length dissipated along the propagation path (experiment #B1). For a color version of this figure, see www.iste.co.uk/richefeu/gravity.zip*

The amounts of energy dissipated over space and time for experiment #B2 are presented for comparison with the results obtained for experiment #B1 in Figures 2.16 and 2.17, respectively. To make the comparison possible in Figure 2.16, and due to the fact that the number of bricks and consequently the amount of total energy are not the same in the two configurations #B1 and #B2, the equivalent energy $W_{eq} \simeq 0.625W$ – where W stands for the work of the forces issued from simulation #B2 – is shown in the right vertical axis. By comparing Figure 2.16 with Figure 2.14 through the equivalent energy W_{eq}, it appears that the amount of energy dissipated by collisions and friction between the bricks along the slope is less in the case of piled bricks. In the transition zone, the total amount of energy together with the energy dissipated by friction at the base of the flow are both lower for piled bricks because of the regularity of the flow resulting from the initial ordered arrangement. In the total event (Figure 2.17), most of the energy is dissipated by friction between the support and the brick (red, 66.8%) and by friction between the bricks (dark orange, 22.4%). The global decrease in the internal friction between bricks during the whole event and the narrowed flow explain the greater runout distance obtained with the avalanche of ordered bricks.

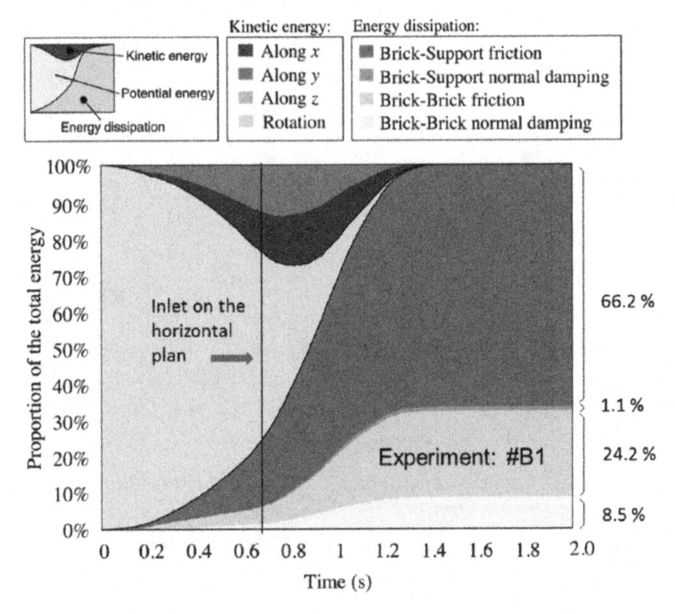

Figure 2.15. *Cumulated dissipated energies as a function of time (experiment #B1). For a color version of this figure, see www.iste.co.uk/richefeu/gravity.zip*

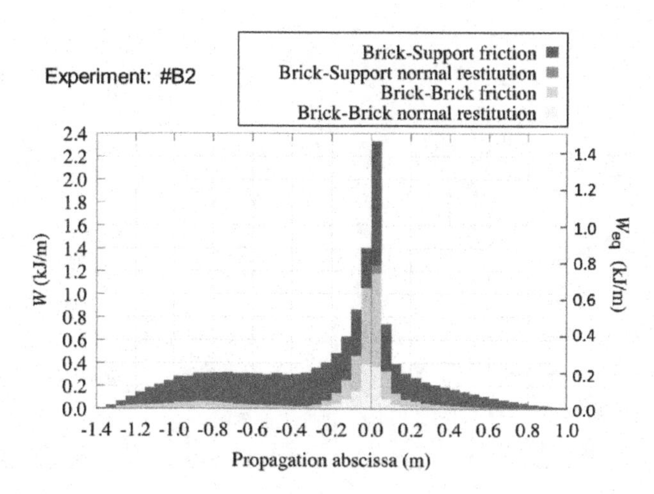

Figure 2.16. *Cumulated energies by unit length dissipated along the propagation path (experiment #B2). For a color version of this figure, see www.iste.co.uk/richefeu/gravity.zip*

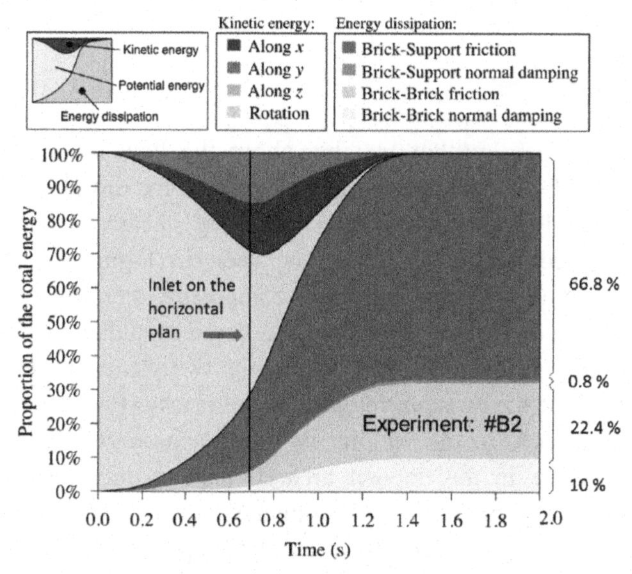

Figure 2.17. *Cumulated dissipated energies as a function of time (experiment #B2). For a color version of this figure, see www.iste.co.uk/richefeu/gravity.zip*

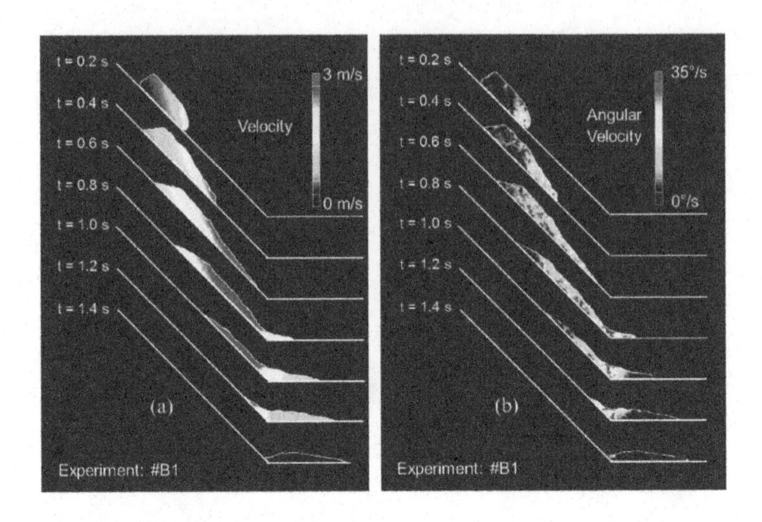

Figure 2.18. *Magnitudes of the translation velocities a) and angular velocities of the bricks b) at various time steps (experiment #B1). For a color version of this figure, see www.iste.co.uk/richefeu/gravity.zip*

2.4.2. *Kinematics*

In order to gain further insights about the flow kinematics, the magnitude fields of translational and angular velocities (Figure 2.18) were computed for configuration #B1 by means of a spatial interpolation technique. As it can be seen in Figure 2.18(a), the velocity magnitude of the particles composing the flow increases regularly as the avalanche develops and declines suddenly when the flow reaches the transition zone between the two planes. Furthermore, the magnitude of the angular velocity of the bricks (Figure 2.18(b)) is much more marked in the vicinity of the transition zone than in the inclined slope or in the deposit area. It appears therefore that the change in direction between the two planes induces a reduction in the velocity magnitude, but it also triggers a broad perturbation of the flow by boosting the particle rotations, which is in accordance with Figure 2.14 showing a greater dissipation in this zone.

The change of the volume over time is shown in Figure 2.19 in terms of the ratio V/V_0 of the actual and initial apparent volumes. The overall mass grows as the flow progresses down the inclined plane until it reaches a maximum of 1.9 – that is an apparent volume almost two times bigger than the initial value. When the avalanche reaches the transition zone at $t = 0.64$ s, the volume starts to reduce as the mass decelerates. It finally stops at an apparent volume 1.4 times bigger than the initial value. These volume changes that can grow or reduce during the flowing and stop phases are of major importance in a real case to properly estimate the area affected by the granular avalanche.

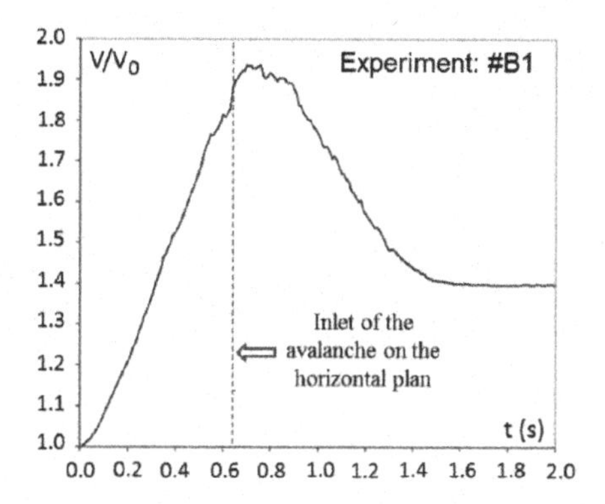

Figure 2.19. *Change of volume over time in the simulation of the experiment #B1*

2.5. Concluding remarks

From the comparisons between the laboratory experiments and the 2D or 3D modeling, it can be concluded that the DEM is able to satisfactorily predict the kinematics of granular flows when the parameters are measured or properly assessed by adequate experimental tests. These comparisons show that, because of multiple interactions of the bricks within the flow, it is possible to satisfactorily approach the main features of the granular deposit (shape and position). As a matter of fact, imperfections in the brick shapes (in particular for the 2D modeling) or uncertainties on the contact parameters have little influence on the collective behavior of flowing bricks. The benefit of the simulations is their ability to give access to micromechanical characteristics or physical mechanisms – that are out of reach with experimental devices – such as the modes of energy dissipation or some specific aspects of the kinematics of the flow. In addition, the use of numerical models makes it possible to easily investigate the sensitivity of any parameter such as the particle shape, the topology of the propagation terrain or the coefficients of energy dissipation.

3

Parameters that May Affect the Flow

It is clear that many parameters have a more or less significant influence on the flow. This chapter aims at describing the role of a selection of parameters related to (1) the amount and the shape of the bricks that constitute the flowing mass, (2) the contact/collision parameters and (3) the roughness of the inclined slope and the "abruptness" of the transition between two slopes.

The work presented in this chapter is the result of collaboration with Guilhem Mollon.

3.1. Constituting blocks

3.1.1. *Amount*

In order to better grasp the influence of the number of particles (or equivalently the size of the particles) on the kinematics of the flow, numerical simulations similar to experiment #B1 (Figure 3.1) were performed using 40 L of randomly poured bricks. The shapes of the bricks were identical in aspect ratios to the original ones [MAN 09], but a homothetic factor was applied to their dimensions. By varying the size of the bricks with a factor ranging from 0.5 to 2, their number was varied from 50,000 to 800 (approximately) as shown in Table 3.1. The analysis carried out here focuses on the shape of the deposits, energy dissipation modes and kinematics of the flows [MOL 15, DAU 15].

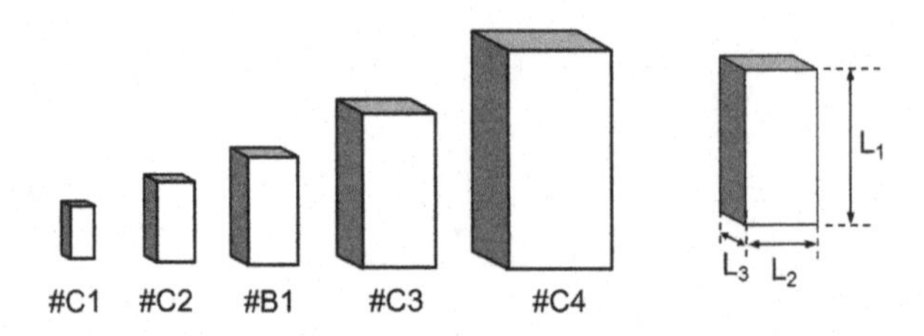

Figure 3.1. *Shape and relative sizes of the bricks used in the simulations*

Simulation	L_1 (mm)	L_2 (mm)	L_3 (mm)	Aspect ratio	Number
#C1	15.50	7.50	4	0.5	49,947
#C2	23.25	11.25	6	0.75	14,868
#B1	31.00	15.00	8	1.0	6,307
#C3	46.50	22.50	12	1.5	1,871
#C4	62.00	30.00	16	2.0	790

Table 3.1. *Size and amount of bricks used in the simulations*

The main characteristics of the deposit morphologies are provided in Table 3.2 and Figures 3.2 and 3.3 for all the simulations performed. To characterize the spreading of the bricks isolated from the main deposit, two statistical variables were defined. The first variable is the "1% fractile in length" defined as the distance along the propagation path and from the slope transition above which exactly 1% of the total number of bricks has been stopped. The second variable is called "1% fractile in width"; it is defined similarly to the previous one except that it reflects the lateral spreading dispersion of the blocks while considering the symmetry of the system. Figure 3.2 is a graphic plot of the data from Table 3.2. It reveals a weak influence of the particle sizes on the Fahrböschung angle, the travel angle and the runout distance. Even if the Fahrböschung and travel angles increase slowly with the size of the bricks, they remain close to 32°and 40°, respectively. The runout and lateral spreading of the bricks are also not greatly affected by the size, since the fractile parameters are rather constant for the simulations #C1 to #C4. On the contrary, the brick size remarkably influences the shape of the deposit: larger bricks (case #C4) causes less extended deposit (both in width and length), while the deposit of small bricks (case #C1) is about 12% more extended with respect to the reference case #B1 (Figure 3.3). Indeed, the deposit resulting from case #C4 seems more "compact" with rather steep slopes on its contour, while the deposit obtained in case #C1 exhibits a planar surface on the top, surrounded by soft slopes.

Simulation	#C1	#C2	#B1	#C3	#C4
Fahrböschung angle (°)	31.16	31.35	31.50	32.37	32.85
Travel angle (°)	39.93	40.04	40.13	40.23	41.15
Runout distance (m)	0.88	0.87	0.84	0.81	0.77
Deposit length (m)	0.95	0.94	0.91	0.87	0.86
Deposit width (m)	1.44	1.41	1.37	1.29	1.16
Deposit thickness (m)	0.10	0.11	0.11	0.13	0.13
1% Fractile in length	0.79	0.79	0.77	0.78	0.73
1% Fractile in width	0.64	0.64	0.63	0.63	0.69

Table 3.2. *Main characteristics of the deposits for the simulations #C1 to #C4*

The variability of the observations was evaluated by performing five additional simulations for each configuration #C2 to #C4. The random arrangement of the bricks in the container was different for each simulation. The mean values and standard deviation of the Fahrböschung and travel angles are shown in Figure 3.4 for each set of simulations. As expected, the standard deviation is less when the number of bricks is increased (i.e. with smaller bricks) but remains limited anyway. From a visual inspection of the deposits (Figure 3.5), the differences are more obvious with large bricks (#C4) than with the small bricks (#C2). However, the greater variability concern in any case the position of isolated bricks, ejected during the flow.

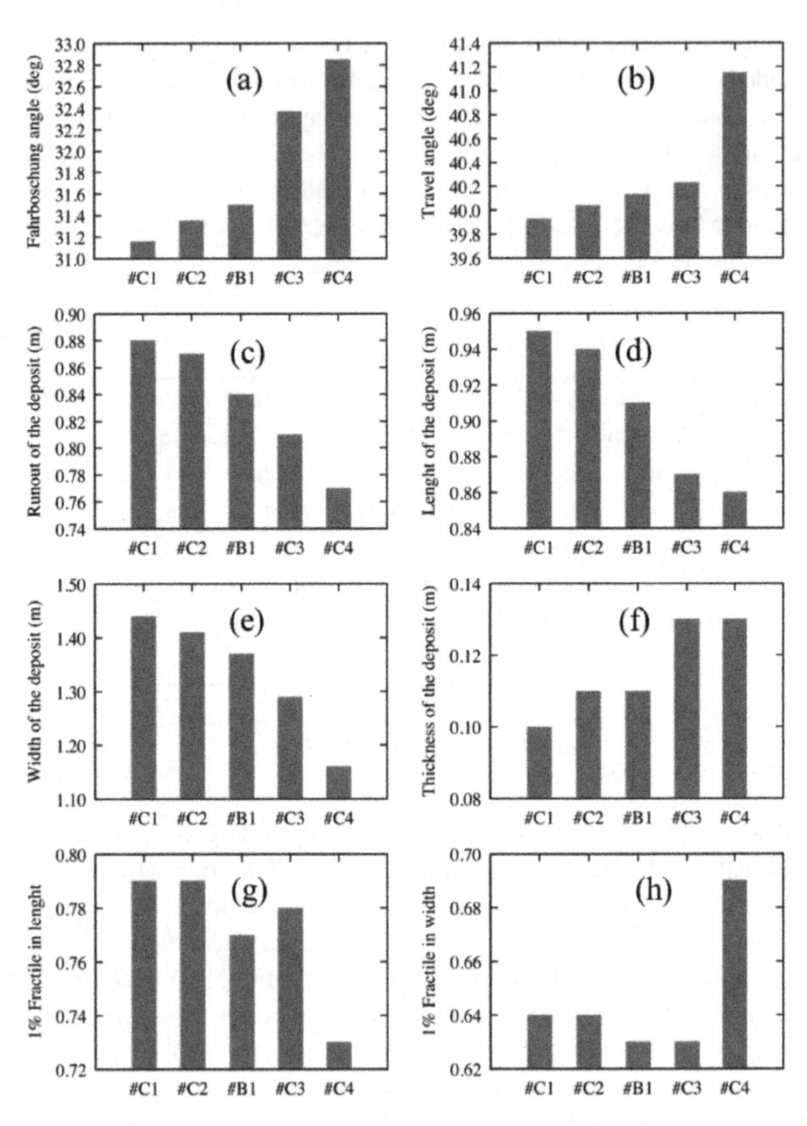

Figure 3.2. *Influence of the size of the particles on the main characteristics of the deposit*

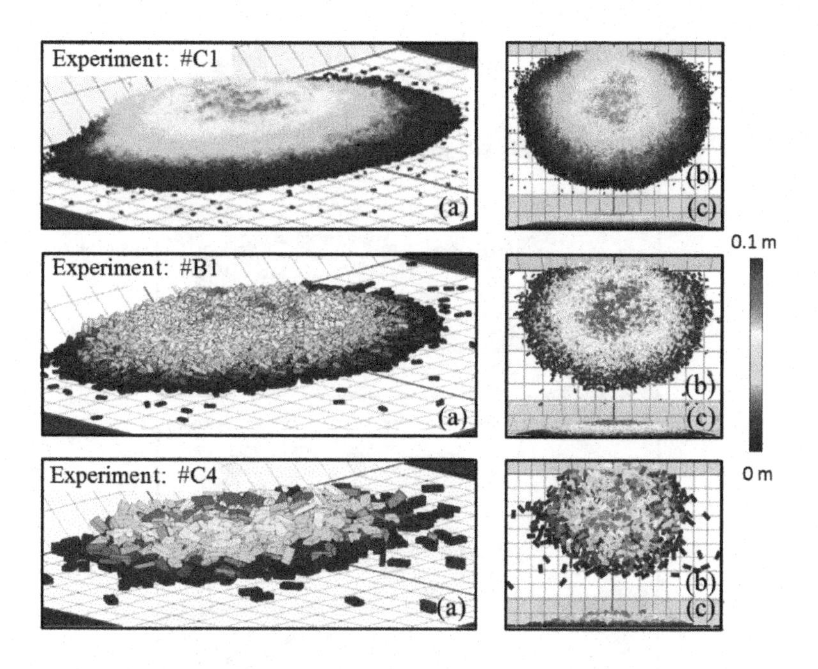

Figure 3.3. *Lateral view a), top view b) and front view (c) of the numerical deposits for cases #C1, #B2 and #C4. For a color version of this figure, see www.iste.co.uk/richefeu/gravity.zip*

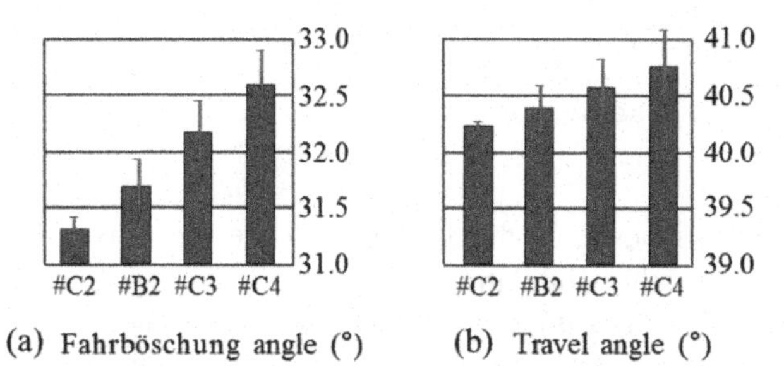

Figure 3.4. *Average values of the Fahrböschung angle a) and the travel angle b) for each set of simulations. The error bars are the standard deviation computed over five simulations*

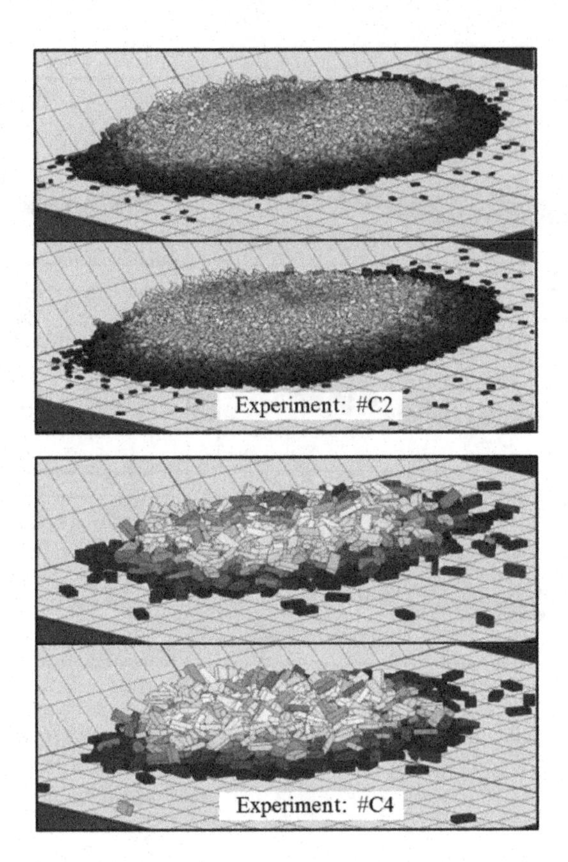

Figure 3.5. *Variability of the final positions of bricks the cases #C2 (small bricks) and #C4 (large bricks). For a color version of this figure, see www.iste.co.uk/richefeu/gravity.zip*

In order to investigate the amount and the location of the energy dissipation for the simulations #C1 to #C4, we have split the propagation path into three zones (Figure 3.6): zone (1) corresponds to the slope, zone (2) corresponds to the horizontal plane and zone (3) extends across the transition line between the two planes. Also, four modes of dissipation are distinguished, namely the dissipation by collisions with the support W_{nBS} (brick–support collisional dissipation), collisions in-between the particles W_{nBB} (brick–brick collisional dissipation), friction with the support W_{tBS} (brick–support frictional dissipation) and friction within the mass W_{tBB} (brick–brick frictional dissipation). The sum of these 12 terms (four modes of dissipation in three different zones) equals the whole dissipated energy, which is constant from one case to another because the total mass of the particles and the initial and final altitudes of the center of mass of the granular material are nearly the same. The dissipative modes are plotted in Figure 3.6 in terms of proportion of the total dissipated energy for the three zones.

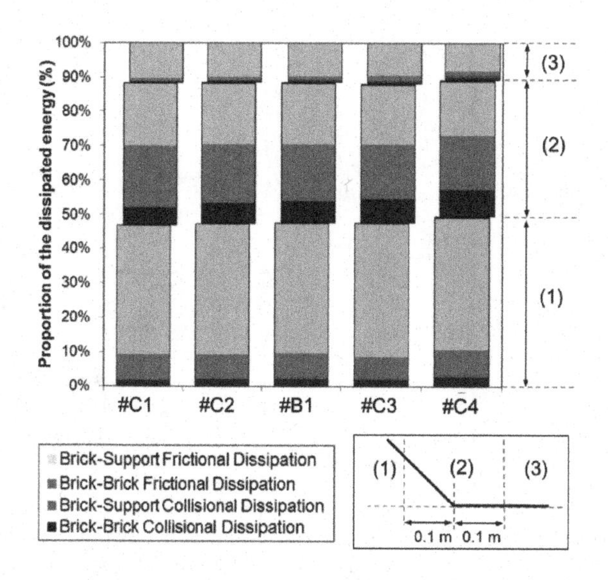

Figure 3.6. *Mode and localization of the cumulative energy dissipations as the function of the block size. For a color version of this figure, see www.iste.co.uk/richefeu/gravity.zip*

Figure 3.6 clearly shows that the main part of the energy is dissipated by friction, mostly between the bricks and the support, which indicates that the modeled avalanches flow mostly in a slipping regime. Only a small fraction of the dissipation occurs by means of collisions, and this dissipation is concentrated in the transition zone for which the flow is disturbed because of the sudden deceleration. After the transition zone, about 10% of the energy remains to be dissipated in zone (3) (i.e. on the horizontal plane). The same trend can be seen in

Figure 3.7 showing the amounts of dissipated energy on each vertical slice along the propagation path (each slice having a width of 0.05 m) for numerical experiments #C1, #B1 and #C4 involving small, medium and large bricks, respectively. Interestingly, the influence of the size of the bricks on the modes and localizations of the dissipations appears to be very limited. This observation is consistent with the fact that the positions of the granular deposits are rather similar regardless the brick size.

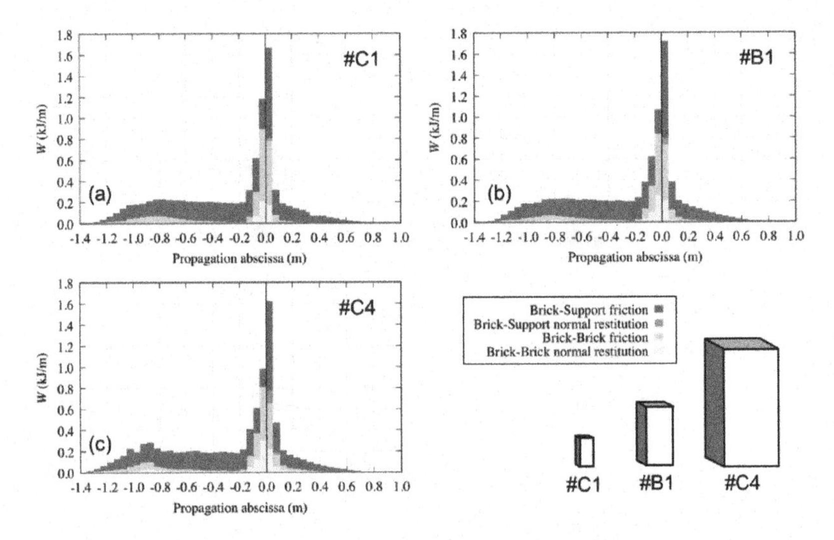

Figure 3.7. *Cumulated energies density (Joules by unit length) dissipated along the propagation path for numerical experiments #C1 (small bricks), #B1 (medium) and #C4 (large bricks). For a color version of this figure, see www.iste.co.uk/richefeu/gravity.zip*

The influence of the size of the bricks on the kinematics of the avalanche, during the propagation phase, is visualized in Figure 3.8 by means of snapshots of the translation and angular velocity fields. These fields are displayed in the longitudinal and a number of transverse cross-sections for the simulations implying smallest and largest bricks, at the instant for which the front of the avalanche reached the abscissa 0.4 m. The velocity profiles are rather similar between the two extreme cases, exhibiting almost homogeneous velocities in vertical cross-sections, with a regular acceleration all along the slope and a rapid deceleration at the vicinity of the transitional zone between the two planes. The main observed differences concern the positions of the bricks in the flow: the small bricks tend to exhibit large lateral spreading during the avalanche, while the large bricks remain more concentrated in the central area. Moreover, along the slope, the angular velocities of the bricks seem to be greater at the surface of the flow for the case #C1. In the transition zone, where the granular flow is strongly disturbed, the rotations of the bricks are also more uniformly distributed in the case #C1 when compared with the case #C4. For the two cases #C1 and #C4, the avalanches are mostly in a slipping regime with homogeneous velocities and limited rotations.

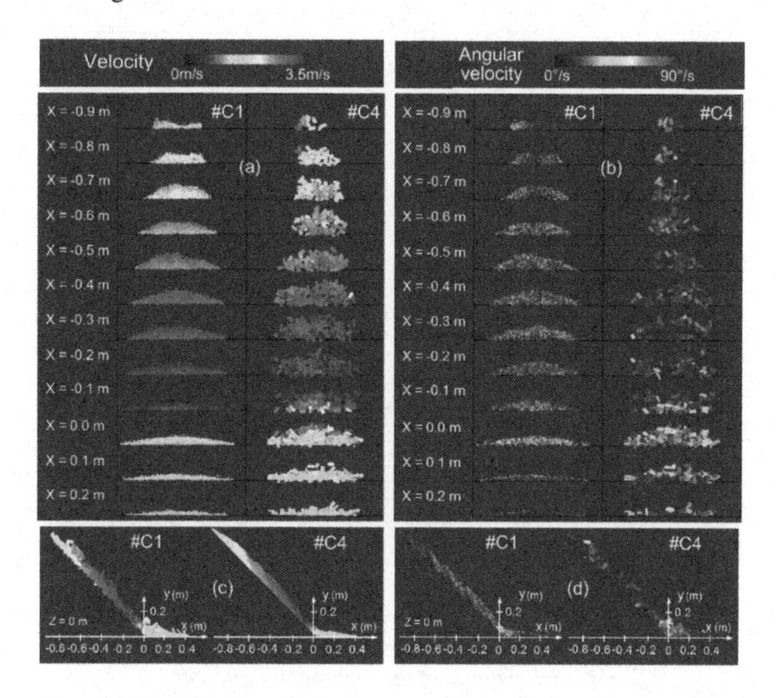

Figure 3.8. *Translation and angular velocities in longitudinal section (c and d) and at several transverse cross-sections (a and b) for the cases #C1 and #C4. For a color version of this figure, see www.iste.co.uk/richefeu/gravity.zip*

3.1.2. *Shapes*

We now focus on the influence of the block shapes on the kinematics of granular flows by varying the block elongation and roundness, randomly packed in a container of 40 L [MOL 15, DAU 15]. For a given release, all the blocks are identical in size and shape. In order to cover a large variety of shapes, rectangular bricks – that is the reference shape – were compared with cubic blocks more or less rounded by "cutting" their corners and thus by altering their sphericity (Figure 3.9). The amount of blocks for each release is almost constant as shown in Table 3.3.

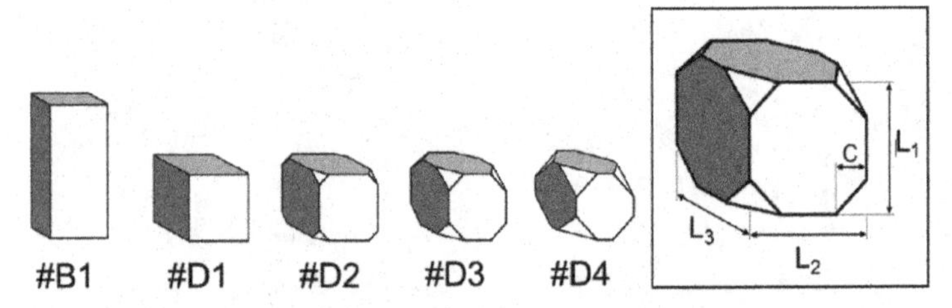

Figure 3.9. *Shapes of the numerical blocks composing the granular material*

Simulation	L_1 (mm)	L_2 (mm)	L_3 (mm)	C (mm)	Number
#B1	31	15	8	–	6,307
#D1	15.5	15.5	15.5	0	6,307
#D2	15.5	15.5	15.5	2	6,318
#D3	15.5	15.5	15.5	4	6,448
#D4	15.5	15.5	15.5	6	6,829

Table 3.3. *Number, size and corner-cutting parameter of the blocks composing the numerical granular materials*

Let us first regard how the block shapes (bricks #B1, perfect cubes #D1 and truncated cubes #D4) affect the position and the morphology of the deposits (Figure 3.10). We observe that the positions of the center of mass of the deposits are quite similar while the amount and the dispersion of isolated particles is increased for perfect cubes, and even more increased for truncated cubes. As a matter of fact, this dispersion is related to the "rolling ability" of these shapes that allows the blocks to roll more, in particular on the horizontal plane after the transition zone. Note also that the perfect cubes (case #D1) are able to arrange locally as a more compact packing (thanks to their regular shape) as perceptible at the vicinity of the bottom of the slope.

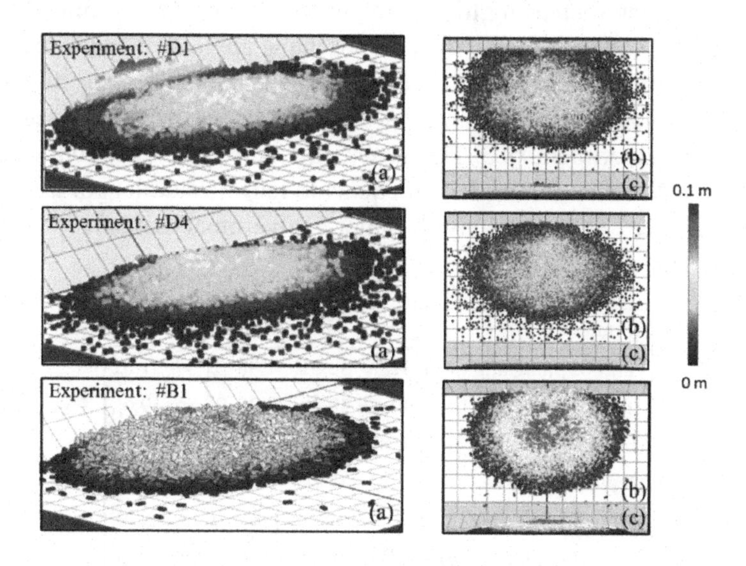

Figure 3.10. *Lateral view a), top view b) and front view c) of the numerical deposits for perfect cubes (case #D1), truncated cubes (case #D4) and bricks (case #B1). For a color version of this figure, see www.iste.co.uk/richefeu/gravity.zip*

Figure 3.11 and Table 3.4 provide the values of the Fahrböschung angle, the travel angle, the runout distance, the dimensions of the deposit and the dispersion fractiles for all the simulations performed. As was mentioned before, an increase in the roundness of the blocks does not seem to have a major effect on the position of the granular deposit (Fahrböschung angle, travel angle and runout distance). The most notable differences can be observed on the 1% fractiles in length and width, which highlight that truncated cubes promote the dispersion of the final deposits. In fact, increasing the sphericity of the particles facilitates their willingness to roll and separate from the main avalanche. The relationship between the particle shape and the ability to roll is very complex and it remains hard to link these two parameters. As an example, when increasing the cutting length C from 0 to 4 mm (i.e. in the cases from #D1 to #D3), we observe a small decrease in the Fahrböschung angle, the length, width and thickness of the deposit, which means that the deposit is more compact. On the contrary, when increasing the cutting length C from 4 to 6 mm (cases #D3 to #D4), the width and the Fahrböschung angle start to reverse the trend and to increase again. It can be deduced from these results that the shape that is more similar to a sphere is the particles used for numerical experiment #C3. This is actually more obvious if you realize that a cube with a cutting length equals to half its side length is, once again, a cube. The sphericity does not evolve monotonically with the parameter C.

Simulation	#B1	#D1	#D2	#D3	#D4
Fahrböschung angle (°)	31.50	32.31	31.88	31.37	31.52
Travel angle (°)	40.13	40.22	40.25	39.85	39.66
Runout distance (m)	0.84	0.76	0.79	0.79	0.79
Deposit length (m)	0.91	0.84	0.84	0.82	0.81
Deposit width (m)	1.37	1.40	1.35	1.33	1.34
Deposit thickness (m)	0.11	0.09	0.09	0.09	0.08
1% Fractile in length	0.77	1.03	1.60	2.11	2.12
1% Fractile in width	0.63	0.77	1.02	1.13	1.03

Table 3.4. *Main characteristics of the deposits for each simulation performed (#D1 to #D4)*

The modes of energy dissipation are presented in Figures 3.12 and 3.13 for bricks (#B1), perfect cubes (#D1) and truncated cubes (#D4). When comparing these different cases, it can be seen that the amount of energy dissipated for the bricks is higher than for cubes (in the zones 1 and 2). This is due to the fact that for cubes, the number of collisions occurring in the bulk increases. Besides, less friction is involved at the base of the flow in the slope (zone 1); the dissipation due to friction in-between the cubic elements is also decreased in the transition up to the mass stop (zone 2 and 3). In fact, bricks have a stronger tendency to slip than cubes, which promotes the dissipation by friction both in the mass and at the base of the flow. Due to the chosen contact parameters in the numerical simulations, the energy dissipation by collisions during the rolling of cubes on the flat zone (zone 3) is higher than the energy dissipated by friction during the sliding of the bricks on the same zone, so that the runout distance of the brick assembly is longer.

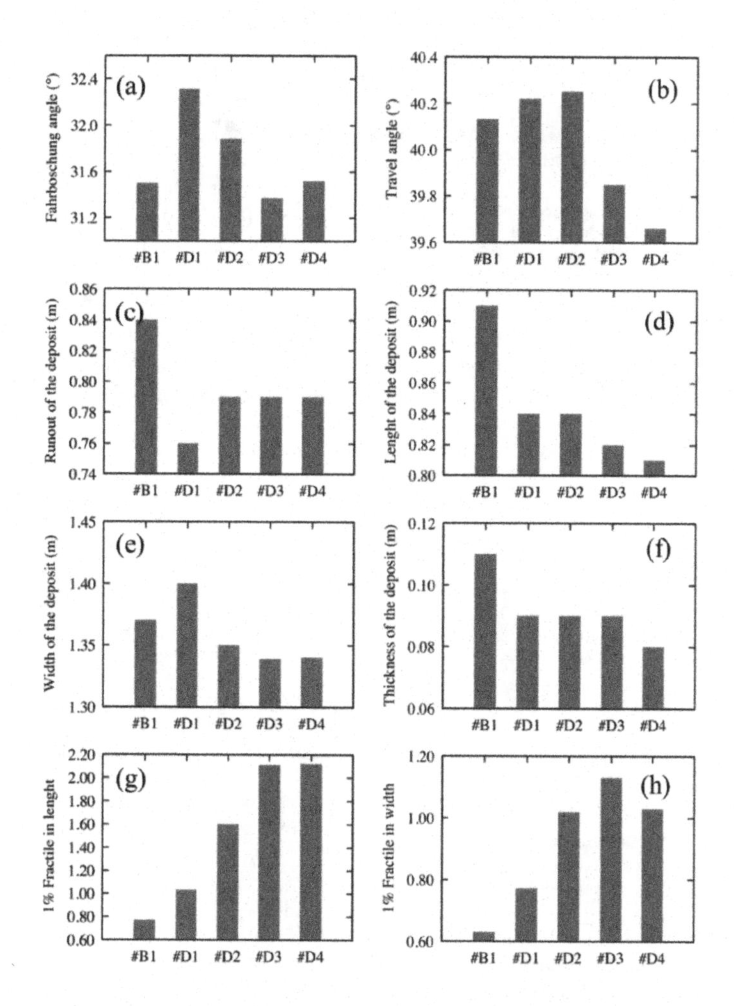

Figure 3.11. *Bar charts of the main characteristics of the deposits as a function of the block shapes (cases #D1 to #D4)*

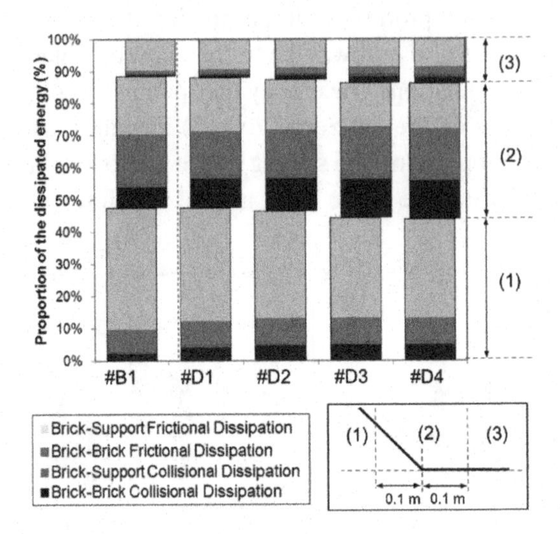

Figure 3.12. *Modes and locations of the cumulative dissipated energies for various block shape (cases #D1 to #D4). For a color version of this figure, see www.iste.co.uk/richefeu/gravity.zip*

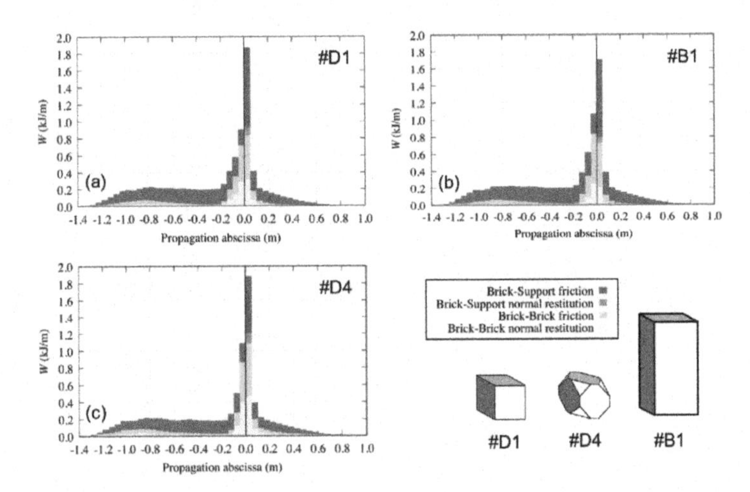

Figure 3.13. *Cumulated energies by unit length dissipated along the propagation path for experiments #B1 (bricks), #D1 (perfect cubes) and #D4 (truncated cubes). For a color version of this figure, see www.iste.co.uk/richefeu/gravity.zip*

The comparison between the results obtained with perfect and truncated cubes (Figure 3.12) shows that the energy dissipations are slightly influenced by the corner-cutting operation. The most noticeable influence is related to an increase in the collisional energy dissipation in zones 1 and 2, correlated with an increase in the block roundness. This increase in the collisional dissipation is probably related to the willingness to roll of particles with cut corners (that increases the collisions in the area of disturbance of the flow), which is balanced by a slight reduction of the basal friction (also probably related to the willingness to roll).

The translation velocity and angular velocity fields are shown in Figure 3.14 for cases #B1, #D1 and #D4. For all block shapes studied, the color gradients reflects a slipping regime on the inclined part through the homogeneous velocity field in vertical cross-sections and the moderate rotation of the blocks. Around the slope transition, the flow is perturbed causing thereby many interactions within the mass. Depending on the sphericity of the particles, we can note an increase in their angular velocities and a change in the stop kinematics on the horizontal plane that leads to a more or less long runout distance of the granular assembly. In fact, the position of the center of mass of the deposit results from complex interaction mechanisms influenced by the shape of the blocks that favors sliding or rotation, and by the contact parameters that promote different modes of dissipation (by friction or collisions).

3.2. Contact parameters

The main difficulties encountered when applying a numerical tool to real events remain in the choice or the determination of the numerical interaction parameters. Without any precise *in situ* solution to access the contact/collision parameters, a sensitivity analysis of the main parameters is crucial. For this purpose, four dissipation parameters of the contact law (e^2_{nBB}, μ_{BB}, e^2_{nBS} and μ_{BS}) varied in the study proposed by Mollon *et al.* [MOL 12]. The reference case for this study is the experiment #B1 concerning the propagation of randomly poured bricks on a two-side slope. Each of the four main dissipation parameters was changed (one value below and one value above), while the other parameters remained unchanged. Including the initial set of parameters, this leads to nine different sets of parameters, which are provided in Table 3.5.

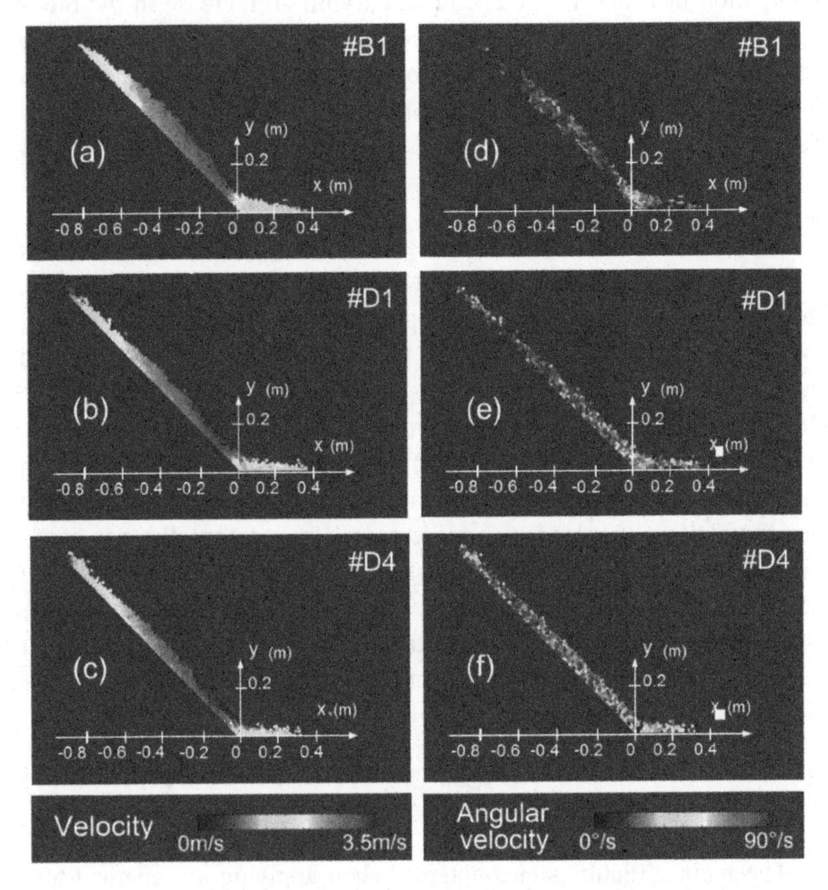

Figure 3.14. *Translation velocities (a, b and c) and angular velocities (d, e and f) in the longitudinal cross-section for experiments #B1 (bricks), #D1 (perfect cubes) and #D4 (truncated cubes). For a color version of this figure, see www.iste.co.uk/richefeu/gravity.zip*

The single parameter that changes compared with the reference simulation (#B1) is in bold.

	#B1	#E1	#E2	#E3	#E4	#E5	#E6	#E7	#E8
e_{nBB}^2	0.13	**0.08**	**0.80**	0.13	0.13	0.13	0.13	0.13	0.13
μ_{BB}	0.86	0.86	0.86	**0.30**	**0.50**	0.86	0.86	0.86	0.86
e_{nBS}^2	0.53	0.53	0.53	0.53	0.53	**0.08**	**0.80**	0.53	0.53
μ_{BS}	0.46	0.46	0.46	0.46	0.46	0.46	0.46	**0.30**	**0.60**

Table 3.5. *Variation of the contact parameters for each numerical case characters.*

The results of nine simulations in terms of runout distance R (m), deposit width W (m), travel angle φ_{CM} (°) and Fahrböschung angle φ_{app} (°) are summarized in Table 3.6 and Figure 3.15. It seems that the contact parameters related to the normal damping e_{nBB}^2 (cases #E1 and #E2) and e_{nBS}^2 (#E5 and #E6) do not have a major influence on the position and shape of the deposits. Moreover, the interblock friction parameter μ_{BB} (cases #E3 and #E4) does not affect notable the position of the deposit, while it influences significantly the shape of the deposit (a decrease in the friction angle between bricks leads to a wider and longer deposit). At the opposite, the basal friction parameter μ_{BS} (cases #E7 and #E8) has a net influence both on the dimensions of the granular deposit and propagation angles.

	#B1	#E1	#E2	#E3	#E4	#E5	#E6	#E7	#E8
Varying parameter	–	e_{nBB}^2	e_{nBB}^2	μ_{BB}	μ_{BB}	e_{nBS}^2	e_{nBS}^2	μ_{BS}	μ_{BS}
Fahrböschung angle (°)	31.86	31.65	31.11	31.26	31.75	31.64	31.44	24.32	35.73
Travel angle (°)	40.24	40.28	39.96	40.06	40.24	40.49	40.24	30.81	45.01
Runout distance (m)	0.82	0.84	0.89	0.88	0.83	0.84	0.86	1.64	0.53
Deposit length (m)	0.89	0.92	0.96	1.09	0.93	0.93	0.93	1.57	0.84
Deposit width (m)	1.39	1.35	1.43	1.65	1.44	1.37	1.35	1.73	1.31
Deposit thickness (m)	0.12	0.12	0.11	0.23	0.11	0.12	0.12	0.08	0.32

Table 3.6. *Main characteristics of the granular deposits for each simulation performed (cases #E1 to #E8)*

The shapes of the final deposits for the numerical experiments #B1, #E7 and #E8 are presented in Figure 3.16. We can see that the lower the basal friction angle is, the more scattered the deposit is;

consequently, the amount of isolated bricks separated from the main deposit is increased.

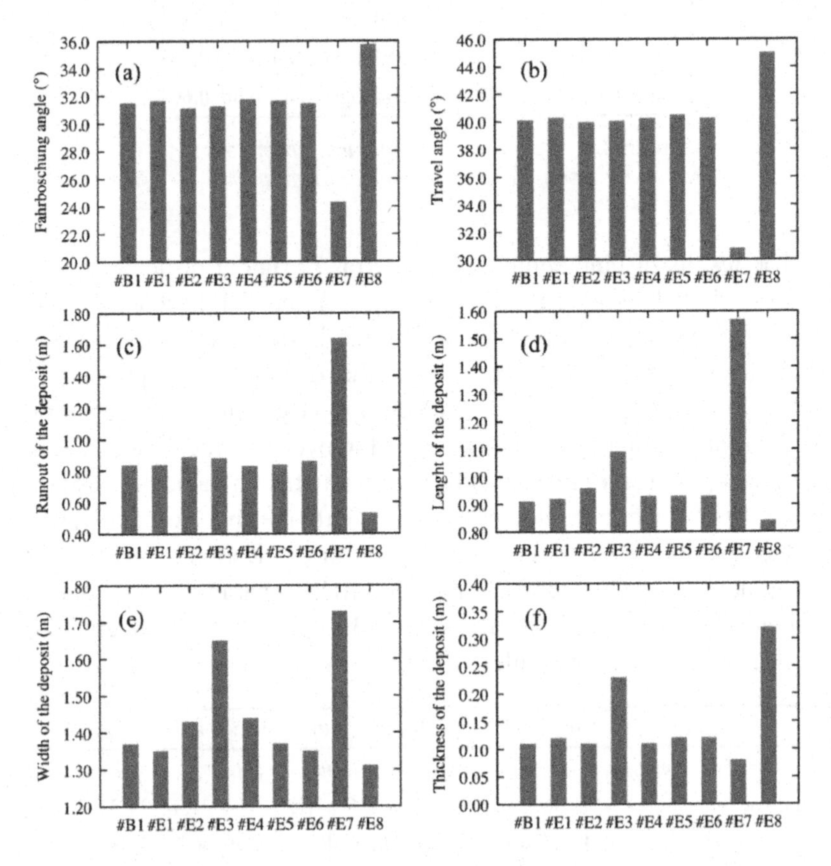

Figure 3.15. *Influence of the dissipation parameters on the main characteristics of the granular deposits (cases #E1 to #E8)*

The amount of energy dissipated along the propagation path for cases #B1, #E7 and #E8 are presented in Figure 3.17. As expected, the amount of energy dissipated by friction along the slope is directly related to the value of the basal friction parameter. For high values of the basal friction, most of the bricks are stopped at the toe of the slope, which explains the large amount of energy dissipated in this zone in the case #E8. Small values of the basal friction induce an increase in the flow velocities and an increase in the intensity of the collisions when the granular flow is abruptly deflected, which explains the peak of dissipated energy observed at the transitional zone for case #E7.

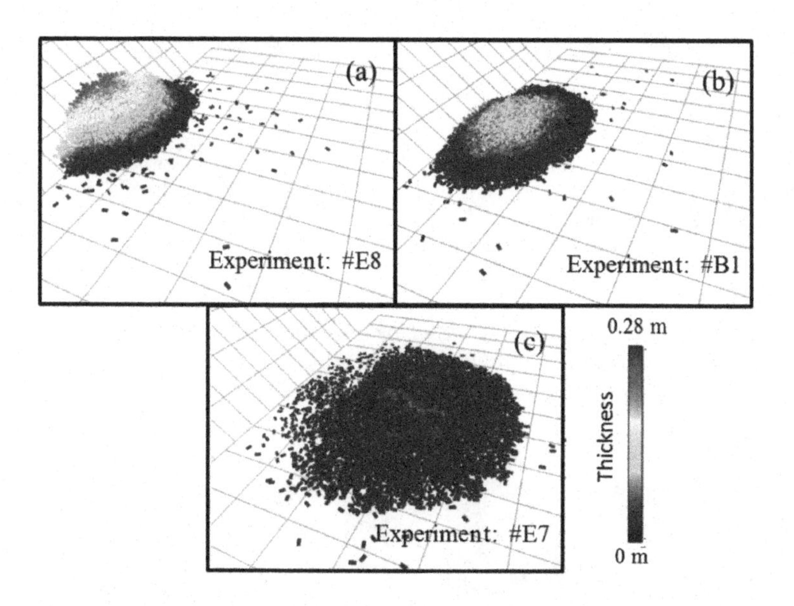

Figure 3.16. *Final brick deposits as a function of the basal friction parameters (Experiments #E8, #B1 and #E7). For a color version of this figure, see www.iste.co.uk/richefeu/gravity.zip*

The total amounts and different modes of energy dissipated during the event can be deduced from Figure 3.18. Contrary to what was expected, the total amount of energy dissipated by friction at the base of the flow increases when the basal friction is decreased: 58.3% of the total amount of energy dissipated by friction at the base of the flow for the numerical experiment #E8 ($\mu_{BS} = 0.6$) and 67.7% of energy dissipated by basal friction in the case #E7 ($\mu_{BS} = 0.3$). This is because small values of the basal friction promote the sliding motions and amplify as a result the amount of energy dissipated by friction at the base of the flow.

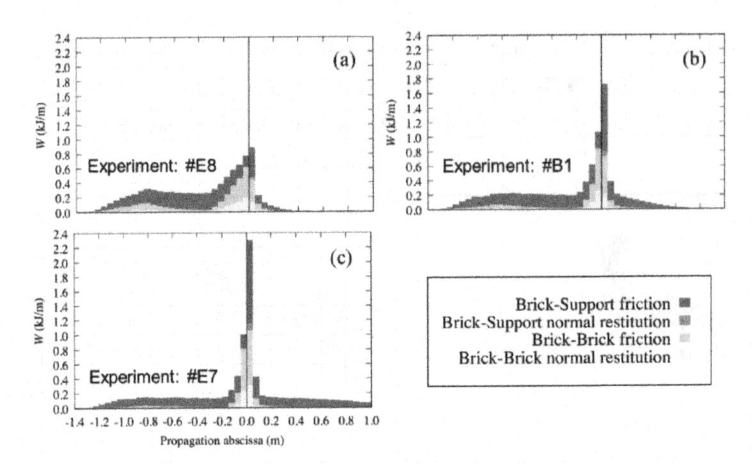

Figure 3.17. *Cumulated energies by unit length dissipated along the propagation path (experiments #E8, #B1 and #E7). For a color version of this figure, see www.iste.co.uk/richefeu/gravity.zip*

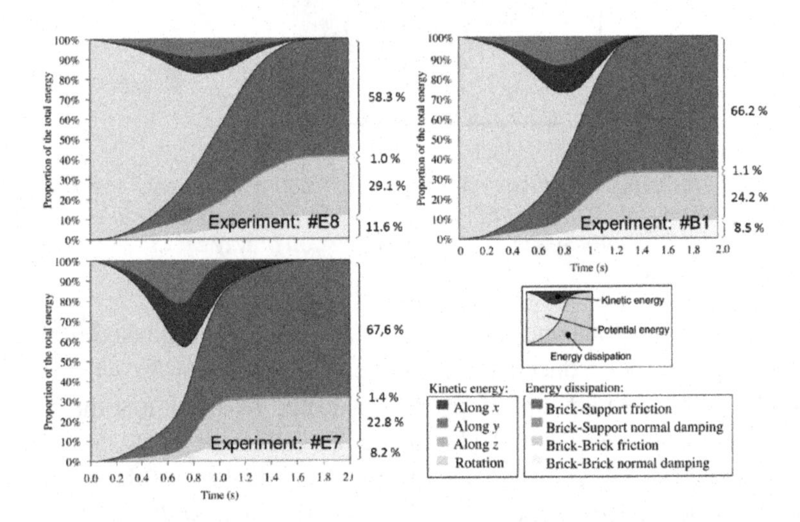

Figure 3.18. *Time course of cumulated dissipated energies (experiments #E8, #B1 and #E7). For a color version of this figure, see www.iste.co.uk/richefeu/gravity.zip*

The conclusion drawn about the main influence of the basal friction parameter is strongly related to the fact that the slope considered in this study is perfectly planar and that the friction coefficient between the blocks is always greater than those used at the base of flow. This condition gives rise to a quite regular and undisturbed flow for which most of the energy dissipation arises by means of basal friction. In the case of a more complex topography, the influences of the normal damping and interparticle friction should probably be much more marked.

3.3. Propagation area

3.3.1. *Abrupt change in slope*

To analyze the influence of the smoothness of the transition between the two planes of the propagation area on the kinematics of the flow, a number of simulations similar to the configuration #B1 (propagation of randomly arranged bricks) were performed by only adding a circular curvature at the transition zone between the two planes [DAU 15]. The curvature radius R_c was varied from zero (no curvature) to nearly 1 m as shown in Table 3.7. To cover a range of initial potential energy, the container was positioned at several heights ($H1 = 1.0, 1.5, 2.0$ or 2.5 m) for simulations #F1 to #F7. The contact parameters, the amount of bricks and their shapes were the same as those used for the numerical experiment #B1.

Simulation	#F1	#F2	#F3	#F4	#F5	#F6	#F7
Curvature radius (m)	0.00	0.05	0.10	0.20	0.40	0.60	0.90

Table 3.7. *Curvature radii of the transition for simulations #F1 to #F7*

The position of the mass center of the deposit is given in Figure 3.19 for all the heights $H1$ and all the curvature radii R_c. We see that the more abrupt the transition between the two planes is (i.e. the smaller the curvature radius is), the shorter the propagation distance of the mass center is. This tendency attests the activation of various dissipative mechanisms in the transitional area. Similar trends are obtained regardless the value of the initial potential energy of the system (i.e. the values of $H1$).

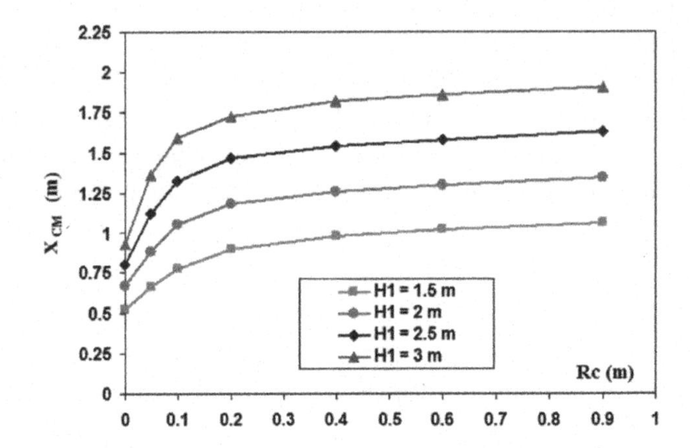

Figure 3.19. *Position of the mass center of the deposit for several heights $H1$ and different values of the curvature radius R_c*

The cumulated energies dissipated along the propagation path are compared for different curvature radii and for an initial height of release $H1 = 2$ m in Figure 3.20. It can be noted that the flow is strongly affected by the smoothness of the transition. Reducing the radius makes the internal dissipation modes (collisions and friction between the bricks) more present; otherwise, the basal friction is the main dissipation way and tends to be the only one with the increase in the radius.

To analyze more precisely the influence of the transition zone on the dissipation modes, the dissipated energies along the propagation path have been plotted in Figure 3.21 for the numerical experiment #F5 ($R_c = 0.4$ m and $H1 = 2$ m). Note that the energies were *not* cumulated this time. We see that the amount of dissipated energy is more significant in the curved part of the propagation way (highlighted in gray in the plot). Everywhere, expected in the curved part, the dissipation by friction with the support is almost constant. On the other hand, in the curved section, the dissipation by friction with the support increased due to the action of the *centripetal acceleration*, which contributes to make the normal forces stronger (and consequently the friction forces also). The total energy dissipated by friction at the base of the flow due to the centripetal acceleration (denoted by W_{ctBS}) is defined as the surface comprised between the W_{tBS} curve and the horizontal dotted line corresponding to the value of the energy dissipated by basal friction before and after the curved transition.

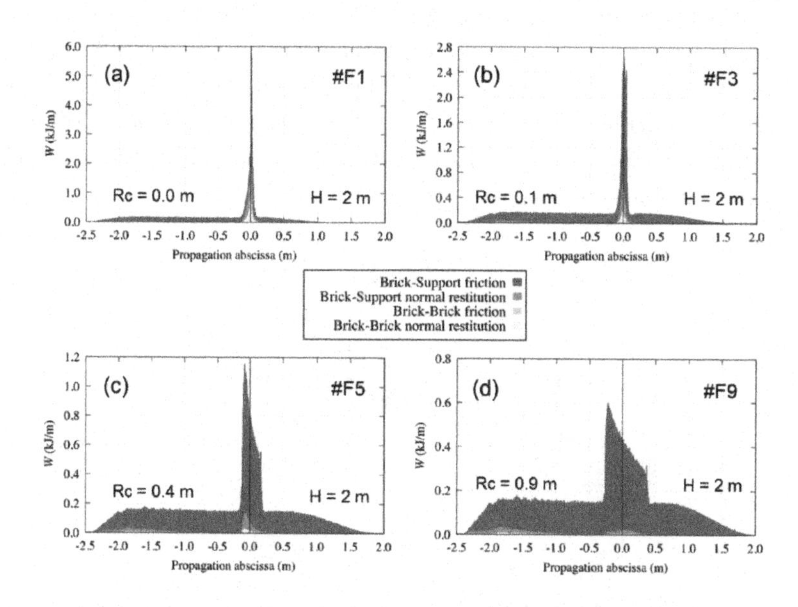

Figure 3.20. *Cumulated energies by unit length dissipated along the propagation path for different values of the curvature radius (R_c = 0, 0.1, 0.4 and 0.9 m) and for the initial release height $H1$ = 2 m. For a color version of this figure, see www.iste.co.uk/richefeu/gravity.zip*

However, it is worth noting that the length of the propagation path changes as a function of the curvature radius. To analyze the exceeding dissipation W_{ctot} that is only due to the presence of curved transition, the latter was split into four dissipation origins in the curved zone: (1) the centripetal acceleration, (2) the collisions at the base of the flow, (3) the friction forces between the blocks, and (4) the block collisions; $W_{ctot} = W_{ctBS} + W_{nBS} + W_{tBB} + W_{nBB}$. As an example, Figure 3.22 shows this splitting as a function of R_c for $H1 = 2$ m. It can be observed that W_{ctot} declines as the curvature radius is increased. Actually, these observations are somehow expected, since increasing the radius makes the transition smoother and consequently makes the dissipation mainly caused by friction at the base of the granular flow. Perhaps more surprisingly, is the fact that the dissipation by friction at the base, resulting from the centripetal acceleration, is almost constant as long as the curvature radius is large enough. At the opposite, for values of R_c smaller than 0.2 m, the induced perturbation of the flow is accompanied by collisional and frictional mechanisms within the granular mass. For an abrupt transition between the two planes ($R_c = 0$), the dissipation by collisions represents about 56% of the total dissipated energy W_{ctot}, while it is less than 6% for the smoothest transitions.

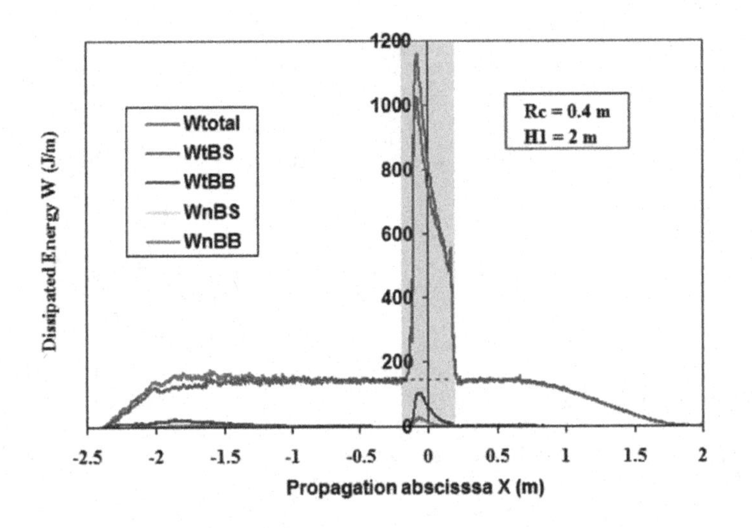

Figure 3.21. *Energies dissipated along the propagation path for R_c = 0.4 m and $H1$ = 2 m (case #F5). For a color version of this figure, see www.iste.co.uk/richefeu/gravity.zip*

The inlet velocity of the blocks at the transition is related to the release height $H1$. Figure 3.23 shows the modes of dissipation as a function of R_c for these different release heights. We observe a linear relation between the energy dissipation W_{ctBS} caused by the centripetal acceleration and the release height regardless of the values of the curvature radius. On the other hand, the energy dissipated within the mass is affected (it increases) by the release height $H1$ only when the flow is strongly perturbed (i.e. for abrupt transition or small values of the curvature radius). Indeed, in the curved section, the contact forces increase in the bulk because of the centripetal acceleration in such a manner that all contact forces increase.

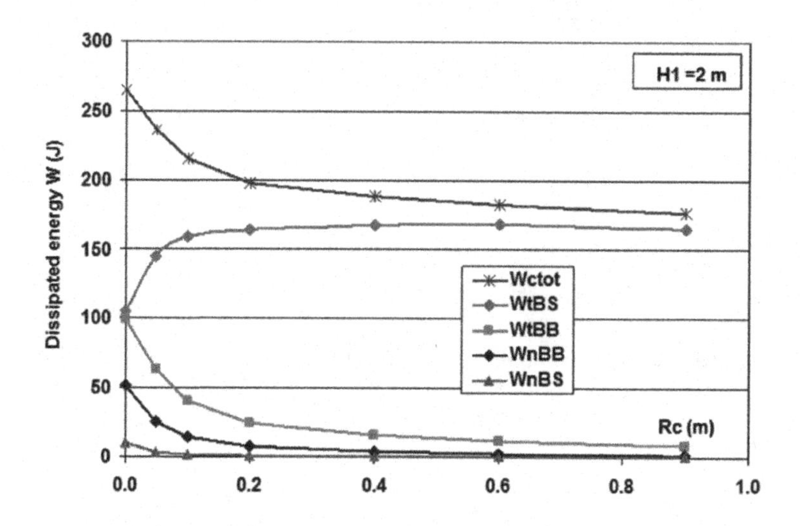

Figure 3.22. *Dissipated energies at the transition zone as a function of the curvature radius (cases #F1 to #E7). For a color version of this figure, see www.iste.co.uk/richefeu/gravity.zip*

Figure 3.24 shows that the energy dissipated in the curved part is either constant or linearly dependent on the height $H1$ of the container (proportional to the squared velocity of the blocks). Presumably, the results presented here are strongly dependent on the thickness of the granular flow that is itself related to the amount of bricks, the shape of the container and its height. These configurations promote or not the sliding mechanisms at the base of the flow and/or the shearing mechanisms within the granular mass. A notable point is that, for the performed numerical experiments, the critical curvature radius below which the flow is not too much perturbed is 0.2 m; this value is six times greater than the length of one brick.

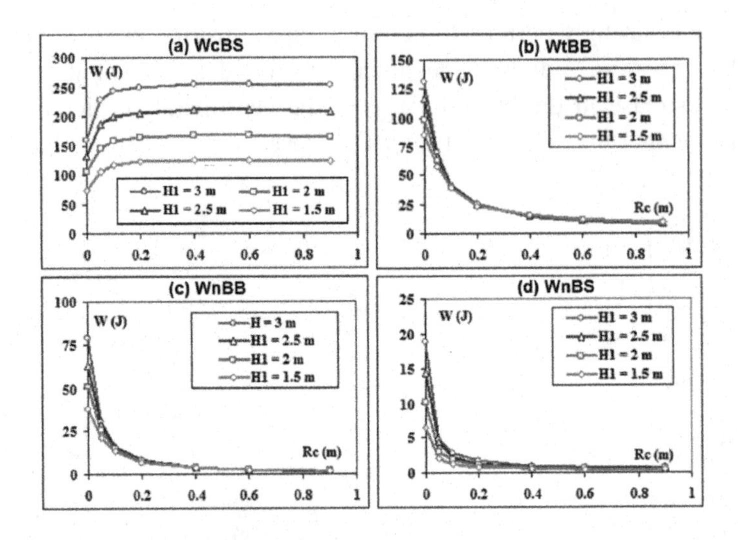

Figure 3.23. *Dissipated energies at the transition zone as a function of the curvature radius for different release heights (cases #F1 to #F7). For a color version of this figure, see www.iste.co.uk/richefeu/gravity.zip*

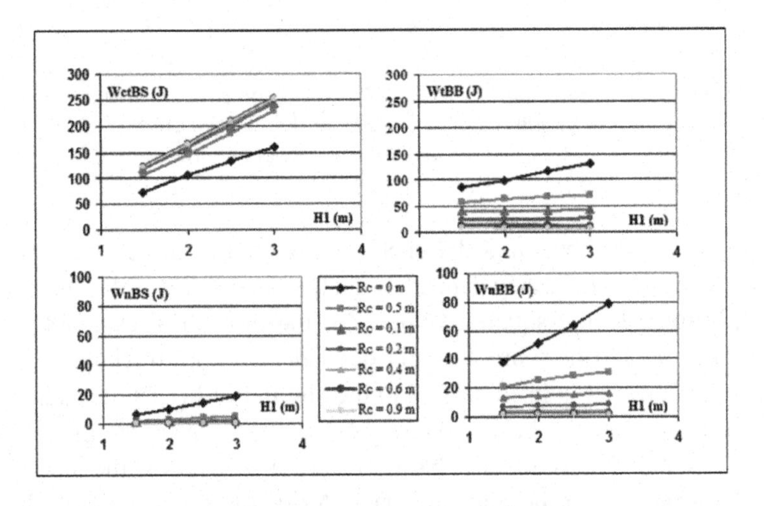

Figure 3.24. *Dissipation at the transition zone for several curvature radii as a function of the release height. For a color version of this figure, see www.iste.co.uk/richefeu/gravity.zip*

3.3.2. *Relative roughness*

When dealing with numerical simulations of real avalanches, particular attention needs to be paid to the definition of the slope geometry. Increasing the resolution of the propagation area to account for the local roughness induces a significant increase in the calculation duration, which is usually not compatible with engineering practices. To determine the influence of the slope roughness on the kinematics of the avalanche (sliding or rotation of the blocks) and on the dissipation modes (by collisions or friction), an additional waviness of the slope was introduced in the form of sharp ripples [MOL 15, DAU 15].

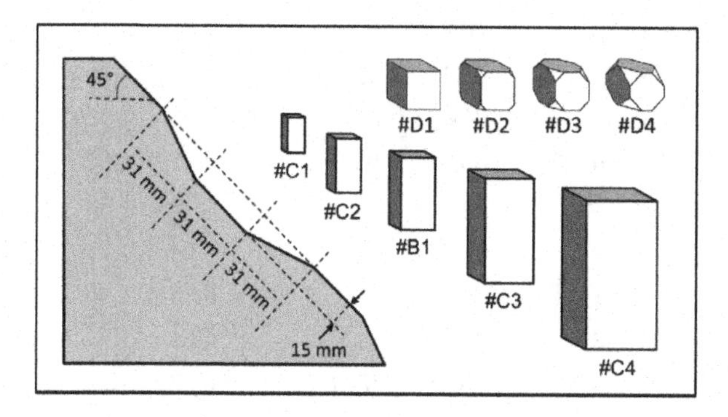

Figure 3.25. *Geometry of the undulations of the slope compared to the brick sizes*

As shown in Figure 3.25, these ripples/undulations are scaled relatively to the dimensions of the original rectangular bricks used in experiment #B1. To evaluate the effects of the block shapes combined with the slope undulation, numerical simulations with small bricks, large bricks and perfect and truncated cubes were performed. The cubes – with or without cut corners – are also smaller than the typical length of an undulation.

The main characteristics of the deposits obtained with undulated slopes are presented, for all block shapes, in Figures 3.26 and 3.27 and compared to those obtained using planar slope in Table 3.8 (the letter "u" refers to the numerical experiments carried out using undulated slopes). Whatever the size of the bricks, the slope waviness makes both the Fahrböschung and the travel angles 4°–6° higher; correspondingly, the runout distances are decreased. Regarding the deposit extensions, the lengths are smaller while the widths are larger.

	Fahrböschung angle (°)	Travel angle (°)	Runout distance (m)	Length (m)	Width (m)	Thickness (m)
#B1	31.50	40.13	0.84	0.91	1.37	0.11
#B1u	35.99	45.52	0.48	0.81	1.48	0.32
#C1	31.16	39.93	0.88	0.95	1.44	0.10
#C2	31.35	40.04	0.87	0.94	1.41	0.11
#C3	32.37	40.23	0.81	0.87	1.29	0.13
#C4	32.85	41.15	0.77	0.86	1.16	0.13
#C1u	35.85	46.09	0.52	0.93	1.52	0.40
#C2u	36.10	45.71	0.50	0.85	1.52	0.35
#C3u	35.48	44.38	0.47	0.75	1.35	0.23
#C4u	35.89	44.19	0.47	0.64	1.23	0.20
#D1	32.31	40.22	0.76	0.84	1.40	0.09
#D2	31.88	40.25	0.79	0.84	1.35	0.09
#D3	31.37	39.85	0.79	0.82	1.33	0.09
#D4	31.52	39.66	0.79	0.81	1.34	0.08
#D1u	34.47	43.97	0.57	0.74	1.54	0.17
#D2u	33.86	43.59	0.56	0.67	1.47	0.13
#D3u	32.93	42.97	0.59	0.67	1.43	0.11
#D4u	33.20	43.13	0.61	0.70	1.47	0.11

Table 3.8. *Main characteristics of the deposits for a selection of simulations with planar slope compared with simulations that use the same configurations except that the slope is undulated*

The observed trend is a direct consequence of the change in how the blocks move (from basal sliding to internal shearing) and in the dissipation modes (from basal friction to bulk dissipation) consecutively to the introduction of the waviness. However, the axial and lateral extensions largely increase with the block size or the angularity of the corners, which means that the waviness of the slope acts differently depending on the shape of particles (small and large bricks, perfect and truncated cubes).

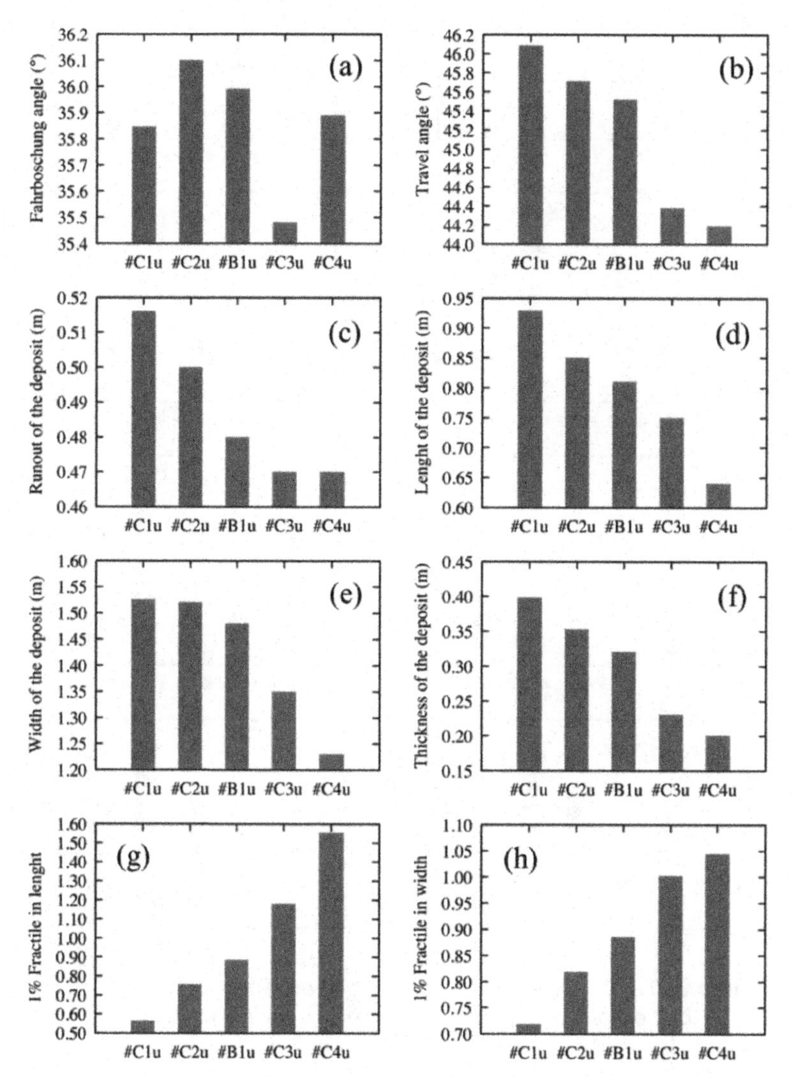

Figure 3.26. *Influence of the brick sizes and undulation of the slope on the main characteristics of the deposits*

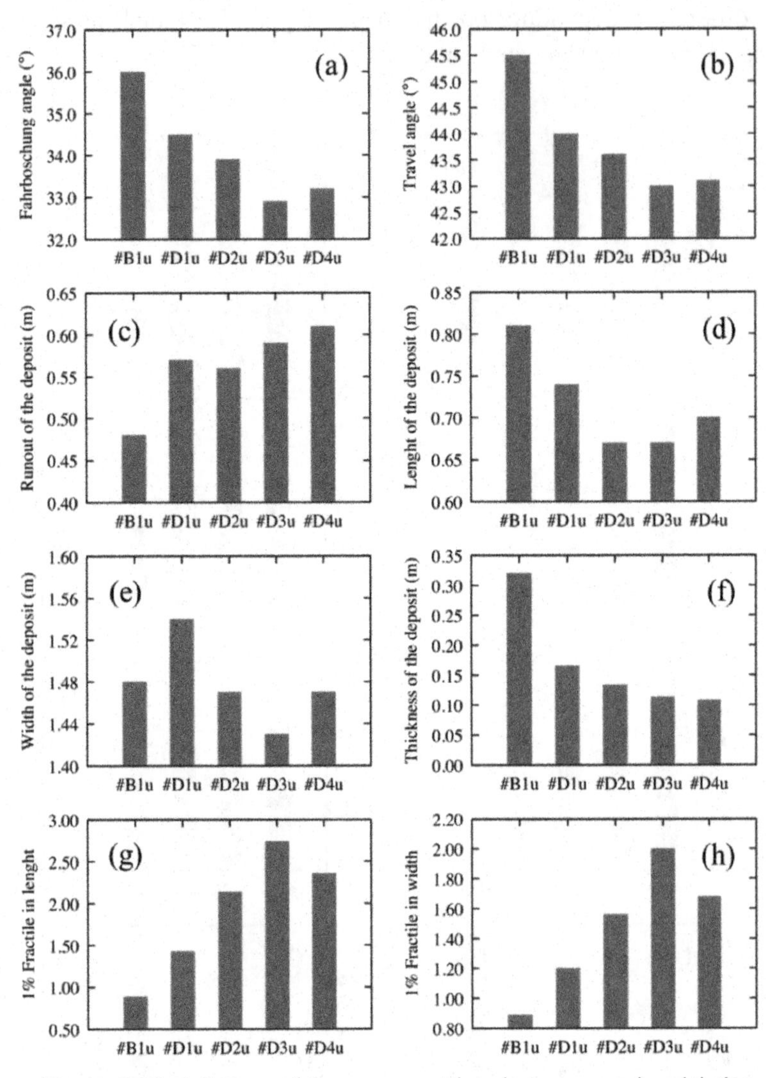

Figure 3.27. *Influence of the corner-cutting deepness and undulation of the slope on the main characteristics of the deposits*

Figure 3.28 and 3.29 provide a comparison of the granular deposits in the four extreme cases (#C1, #C4, #D1 and #D4) both for planar and undulated slopes. The amount of particles in rest at the bottom of the inclined slope is systematically greater in the case of undulated slopes, regardless the shape of blocks. At the opposite, the particles scattered far from the main deposit in each direction are always more numerous with undulated slopes. In fact, the waviness of the slope perturbs the flow, which causes more dissipation within the mass and therefore smaller runout distances. The numerous collisions and excessive rolling of the blocks gives rise to ejections (the more spherical the blocks are, the longer the distance from the main deposit is).

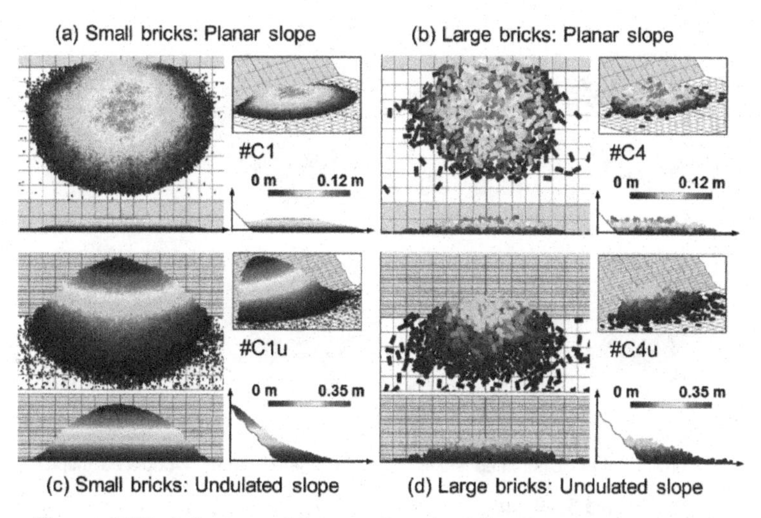

Figure 3.28. *Influence of the roughness of the slope on the granular deposit in the case of small (case #C1) and large (case #C4) bricks. For a color version of this figure, see www.iste.co.uk/richefeu/gravity.zip*

The comparison between planar and undulated slopes of the dissipation modes along the slope is given in Figure 3.30 for all tested block shapes. As seen previously, a difference is observed between the dissipation on the slope (zone 1), in the transition zone (zone 2) and on the horizontal plane (zone 3). The dissipation modes are rather different according to the roughness of the slopes, regardless of the shape and size of particles. Clearly, the energy dissipated on the slope by friction or collisions within the mass increases greatly, while the frictional dissipations at the base of the flow are strongly decreased. In the transitional zone (zone 2), the dissipation within the mass is rather constant while frictional dissipation also decreases. In the case of an undulated slope, when comparing the influence of the brick sizes relative to the wave size, a greater dissipation along the inclined slope can be observed for small bricks (case #C1) than large bricks (case #C4). At the opposite, truncated cubes (case #D4), which are more willing to roll than perfect cubes (case #D1), dissipate less energy on the inclined slope. Beyond a certain value, the cutting parameter seems to have less influence on the dissipative modes.

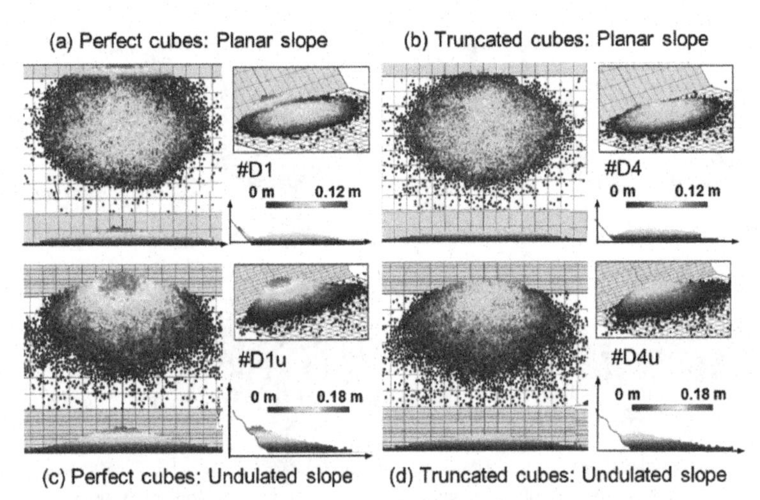

Figure 3.29. *Influence of the roughness of the slope on the granular deposit in the case of perfect (case #D1) and truncated (case #C4) cubes. For a color version of this figure, see www.iste.co.uk/richefeu/gravity.zip*

To point out the influence of the waviness of the slope on the kinematics of the flow, we present in Figures 3.31 and 3.32 the amounts of energy dissipated along the propagation path for all the block shapes tested. These figures clearly show that the dissipation modes are directly influenced by the roughness of the slope. By comparison with the results from Figure 3.31, it can be seen that the main dissipation arises in the bulk for small bricks, while large bricks have a higher tendency to develop friction with the slope. As excepted, at the beginning of the event where sliding mechanisms are favored, the disturbance of the flow results from numerous interactions between the bricks or between the bricks and the upper part of the ripples in the slope, regardless the brick sizes. Figure 3.32 shows that the willingness to roll reduces the dissipation on the bumpy slope. The energy dissipated by friction with the slope is also reduced.

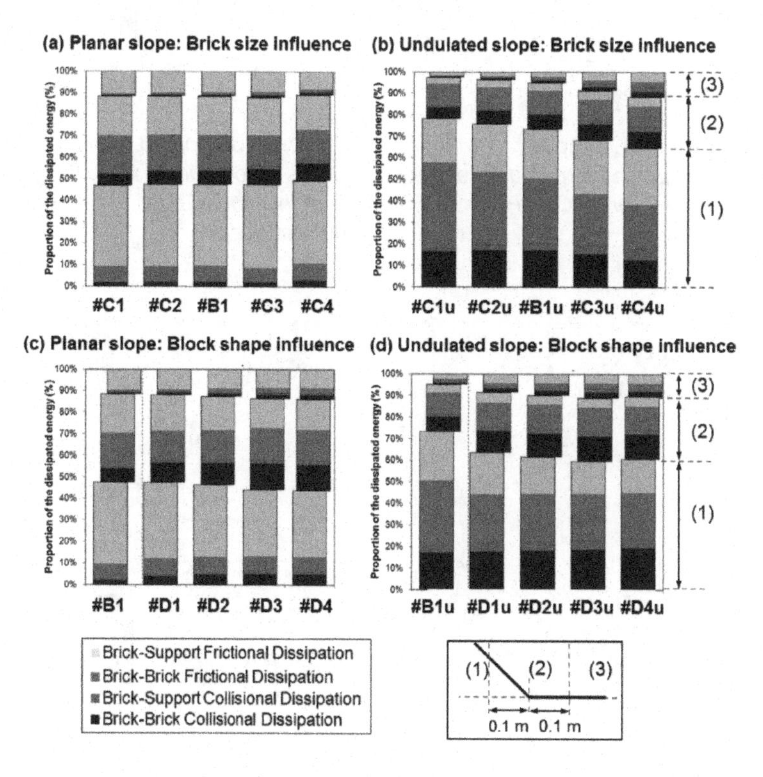

Figure 3.30. *Influence of the undulations of the slope on the location and mode of energy dissipations for various block geometries. For a color version of this figure, see www.iste.co.uk/richefeu/gravity.zip*

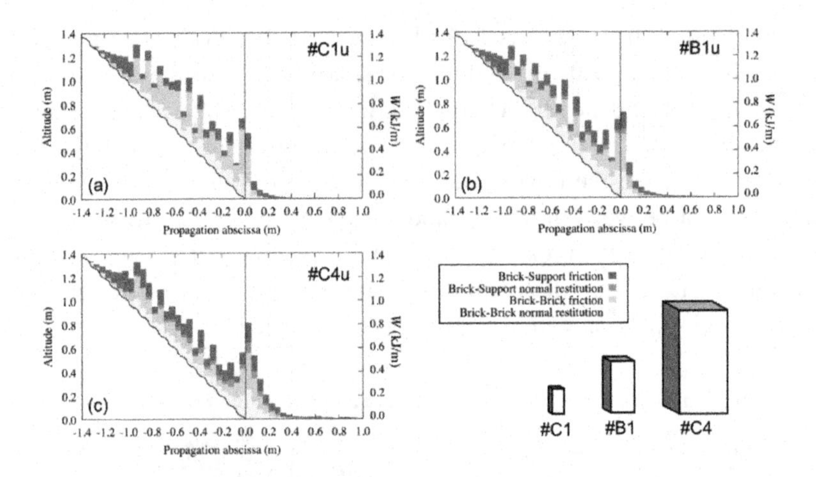

Figure 3.31. *Cumulated energies by unit length dissipated along the propagation path for different sizes of bricks (experiments #C1, #B1 and #C4). For a color version of this figure, see www.iste.co.uk/richefeu/gravity.zip*

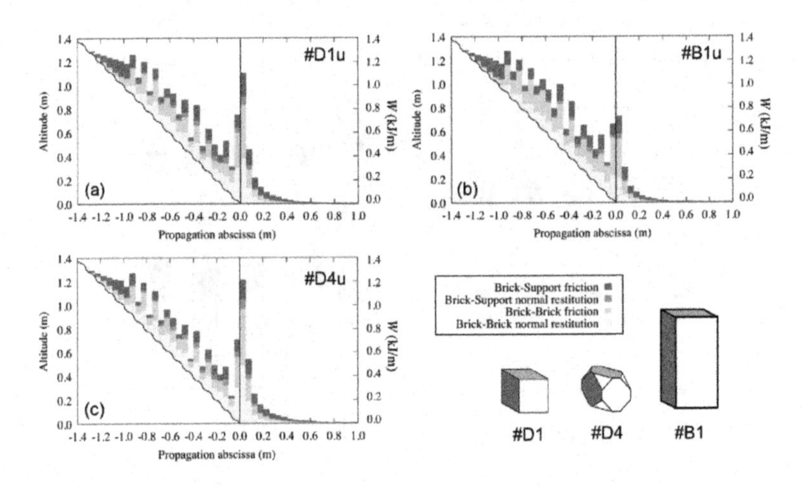

Figure 3.32. *Cumulated energies by unit length dissipated along the propagation path for different shapes of blocks (experiments #D1, #B1 and #D4). For a color version of this figure, see www.iste.co.uk/richefeu/gravity.zip*

The time evolution of the four kinds of cumulative energy dissipations for various block sizes, on a planar and an undulated slope, are summarized in Figure 3.33. In these figures, the blue arrows point from the smallest blocks (case #C1) to the biggest blocks (case #C4). The size of blocks induces some trends that are often the same on a planar or an undulated slope. More collisional dissipation arises for larger blocks (either in the bulk or with the support), while the frictional dissipation is, on the contrary, reduced with the growth of blocks – the brick–support frictional dissipation is however an exception since it remains almost constant. Hence, whatever the type of slope, larger bricks flow with a more collisional regime, while smaller bricks promote the internal shearing of the material. Quantitatively, the block size plays a less significant role on a planar slope than on an undulated slope.

As demonstrated above, when flowing on a bumpy terrain, more dissipation arises as a result of the induced disturbance: the undulations allow for higher relative velocities between the blocks, both in tangential and normal contact directions (Figure 3.34). For small blocks (#C1 or #D1), a very clear vertical gradient of the velocity appears on the longitudinal cross-section, with interesting velocity patterns exhibiting the same periodicity than the slope undulation. The free surface of the granular mass has a much higher velocity than the bottom one, which is submitted to periodical cycles of acceleration–deceleration depending on the local inclination of the slope. The transverse cross-sections show that the velocity profile is rather homogeneous in the entire width of the granular mass, since the velocity magnitude is almost independent on the position of the section. When considering large blocks (#C4 or #D4), this stratified flow is not observed. Indeed, in these cases, we observe no lower layer with very limited velocity like in the case of small bricks, and consequently the velocity profile is almost constant in any transverse cross-section. Similar trends can be observed for the angular velocities. In particular, large blocks induce smaller angular velocities – except in the transition zone – while the flow with small blocks exhibit periodic patterns of high angular velocities that alternates with zones of more calm regime (related to the wavelength of the undulations of the slope).

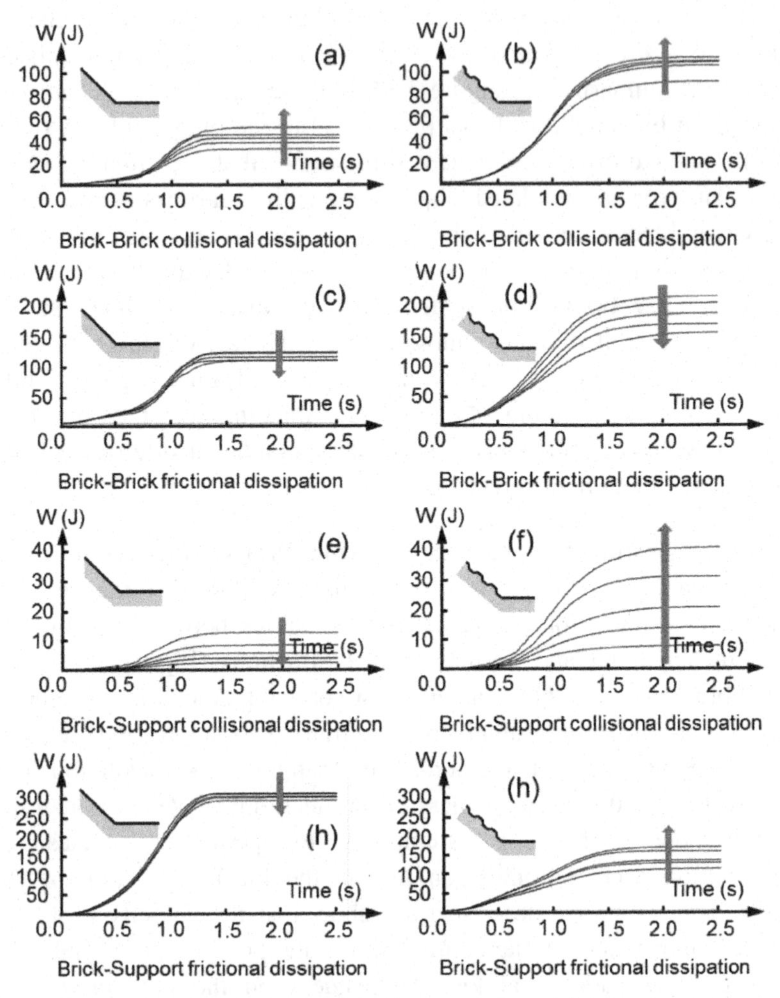

Figure 3.33. *Comparison of the time evolution of the four types of energy dissipations for various block sizes on a planar and an undulated slope*

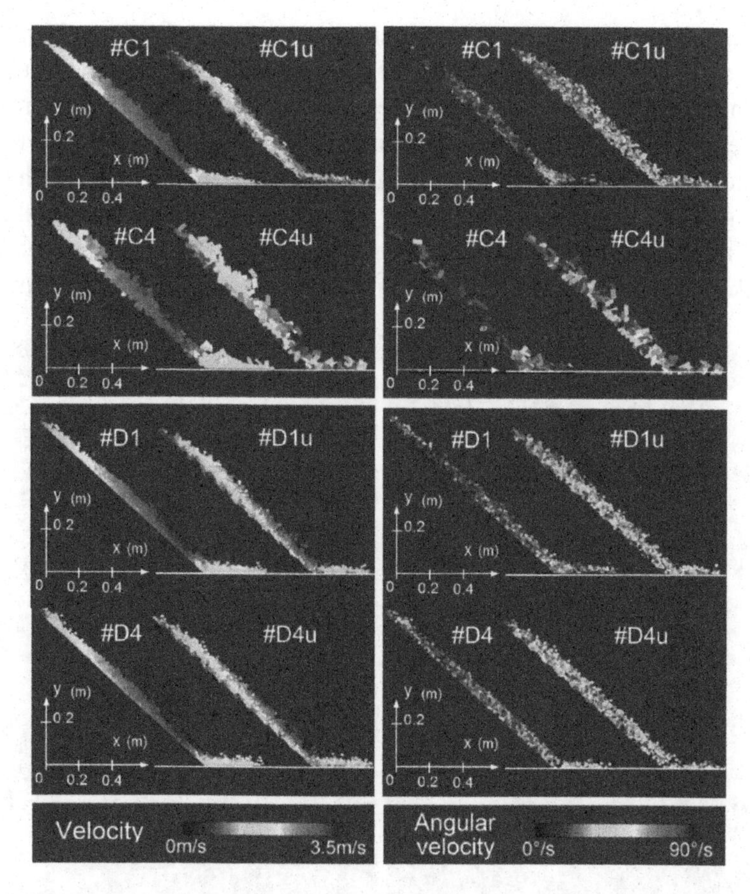

Figure 3.34. *Influence of the undulations of the slope on the translational and angular velocities in longitudinal section and several transverse cross-sections for several block shapes. For a color version of this figure, see www.iste.co.uk/richefeu/gravity.zip*

Figure 3.35 presents the results in terms of local solid fractions for small and large bricks. On a planar slope, the density of the granular packing is rather homogeneous in the entire avalanche (except at its ends), but the avalanche of small blocks is denser than the avalanche of large blocks. The case of large bricks on an undulated slope also highlights a rather homogeneous density, but the granular packing is very loose in this case. The case of small bricks on an undulated slope is the only one for which there is a heterogeneous distribution of the solid fraction. In this simulation, periodic patterns of density follow the undulations of the slope; the higher densities are located in the troughs of the slope and lower densities are rather on the ridges. These density patterns arise at the same places than those observed in the velocity field (Figure 3.34), and the areas of high density are well correlated with the areas of low velocity. Moreover, the transverse cross-sections also show that the density is not homogeneous in the width of the avalanche: the packing is globally denser in the central part than in the lateral parts.

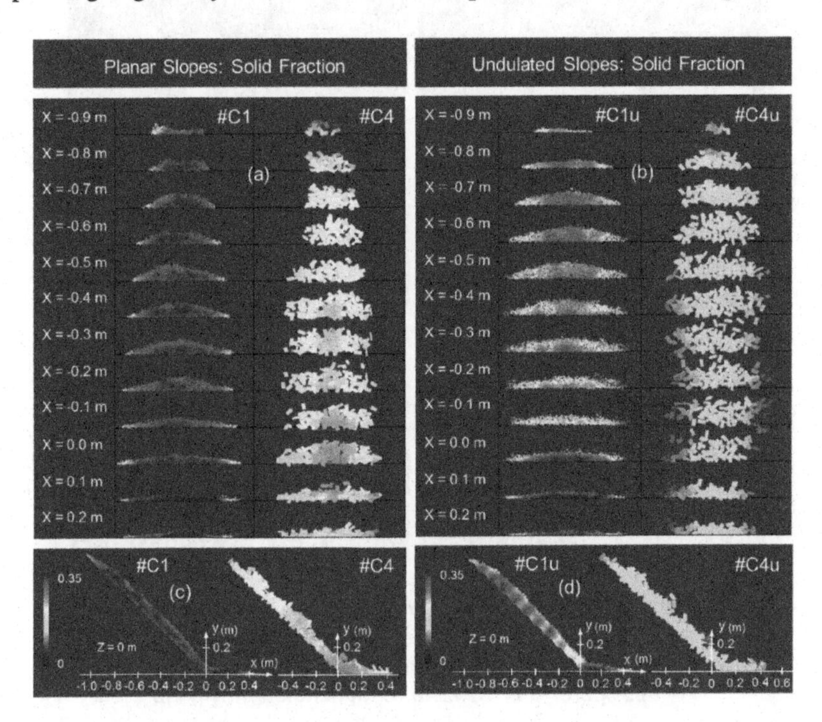

Figure 3.35. *Influence of the undulations on the solid fraction of the granular material for several transverse cross-sections and for smallest and larger blocks. For a color version of this figure, see www.iste.co.uk/richefeu/gravity.zip*

3.4. Concluding remarks

This study, conducted to point out the influence of the main parameters, revealed complex linkages between the geometric properties of the blocks composing the avalanche (size, aspect ratio, sphericity), the dissipation parameters and the topology of the slope (waviness and sharp change in inclination). All these connected features play a complicated role in the prediction of the final morphology and position of the granular deposit and global dispersion of the blocks separated from the main mass. The block shapes and the topology of the slope seem to be essential in the kinematics of an avalanche while the dissipation parameters have a direct influence on the dissipative modes and on the runout distance and position of the deposits.

On a smooth planar slope, the avalanche is mostly in a translational regime (except in the transition zone, where the flow is quite disturbed), the velocity field is constant in transverse cross-sections and the particle rotations are limited. Most of the kinetic energy is dissipated by friction at the base of the flow so that the most influential parameter on the position of the deposit is the basal friction angle. The shape of the blocks has a weak influence on the position of the center of mass of the deposit but a greater influence on its shape and on the amount of scattered blocks outside the main deposit.

A progressive change in inclination of the slope has limited influence on the kinematics of the flow, while an abrupt change disturbs the flow and causes strong shearing in the bulk, marked velocity gradient and numerous block rotations. As a result of the shearing within the granular mass and block rotations, most of the energy is dissipated by friction and collisions within the mass, which limits its propagation. Depending of the block shapes, this change in the flow kinematics activates different modes of dissipation (sliding or rolling) that affect both the position and the shape of the deposit, and the surrounding scattered blocks.

On an undulated slope, the ripples make the flow regime more collisional and reduce the frictional dissipation on the slope and in the transition zone so that the travel angle is increased and a large amount of particles stay at the base of the slope (before the transition). With bumpy slopes, the characteristics of the avalanche are strongly affected by the size of the blocks. Block sizes smaller than the characteristic scales of the slope waves induce strong internal shearing, sharper vertical velocity gradient, higher angular velocities and a quite dense packing. In this case, velocities, angular velocities and local solid fractions are very well spatially correlated with the slope undulations, leading to periodic heterogeneities. On the contrary, block sizes larger than those of the slope ripples make the flowing regime more collisional with fluctuating velocity between neighboring particles, higher angular velocities and looser packing, but without apparent structure. This kind of "turbulence" results in a reduction of basal friction, which generates avalanches with higher mobility and more scattering of isolated blocks.

When dealing with real applications, particular attention needs to be paid to determine the right sizes and the realistic shapes of the blocks that may be involved in the avalanches. The resolution of the propagation area should be of the same order than the block sizes to catch correctly the roughness of the slope and the proper kinematics of the flow. The survey conducted here confirms that the contact parameters need to be estimated with care. Back analysis of past events may be of interest to experiment the whole set of parameters.

4

Application to Actual Rockfalls

This chapter aims to present real-life studies using the code DEMbox. Each of the three cases presented was chosen for their remarkable specificities. The first case is a naturally triggered event that involves relatively few blocks – rather big and elongated – in which the propagation path is quite short. The second case is an artificially triggered event by means of explosive charges. The size grading of the blocks is consequently narrow. The propagation path is long with a soft ground in certain regions. The third case is an attempt to use the model as a predictive tool.

The work presented in this chapter is the result of collaboration with Dominique DAUDON, Yeison-Stiven CUERVO and IMSRN (www.imsrm.com).

4.1. Retro analysis of a natural rockfall implying a few blocks

4.1.1. *Purpose of the modeling*

This application deals with the retroanalysis of a natural rockfall of approximately 1,000 m^3 for which the number and size of blocks are well known and the geometry of the propagation slope is quite planar and well established [CUE 15]. Since the site was easily accessible, it has been possible (1) to precisely localize the position of the blocks on the slope after the event, (2) to characterize the fracture network and (3) to obtain with high accuracy the cliff geometry and the topography of the rocky slope via a photogrammetric campaign.

The parameters of the contact law were initially estimated from orders of magnitude generally recognized in the literature and then adapted slightly as a result of a sensitivity analysis based on the best match between the experimental and numerical positions of the block deposits [CUE 15]. Due to the peculiar shape of the blocks (elongated cuboid blocks), it was not necessary to introduce rolling resistance between the rocky slope and the blocks, so that the block/soil interaction only accounts for the energy dissipated by collisions and friction/abutment at each collision. The contact law parameters are:

– the coefficient of normal energy restitution e_n^2;

– the tangential dissipation coefficient μ that accounts for both Coulomb friction and buttress forces taking place in the case of large penetration of a block in the substrate;

– the normal and tangential contact stiffness (k_n and k_t, respectively).

Besides, the cohesion between blocks taking place in the fracture network was not modeled, this means that the fracture energy was neglected compared to the kinematic energy developed by the free fall motion.

4.1.2. *Description of the event*

The rockfall event took place in 2007 close to the city of Millau in the south of the Massif Central Mountains in France. About 1,000 m^3 of limestone benches, approximately one meter thick, were separated from the quasi-vertical cliff by sliding without toppling. The blocks were scattered on a natural soil with a slope of 30.5° and they were then stopped on a nearly horizontal surface (the inclination angle is less than 6°; Figure 4.1). The propagation path was 56 m long in the inclined slope, and approximately 15 m long on the last quasi-horizontal part.

Most of the rocks were deposited in a 50 m long and 40 m wide zone, and some isolated blocks were ejected far (~70 m) from the cliff. 30 blocks, of volume ranging from 10 to 36 m^3, were identified inside the main deposit and spotted using a GPS tool. These blocks represent 85% of the total mass volume and reflect the rockfall deposit limits. Approximately 10 rock blocks have very large volumes (between 33 m^3 and 36 m^3) and rather similar sizes (6 m length, 3.5 m width and thickness ranging between 1 and 1.5 m) (Figure 4.2).

Figure 4.1. *Overall a) and detail b) views of the cliff after the event, view of the rocky propagation slope c) and geometric characteristics of the slope and vertical cliff*

Horizontal and vertical failure networks were identified at the external surface of the cliff. Horizontal cracks are regularly spaced every meter or every 1.5 m, which explains the variation in the thickness of the blocks of the deposit (Figure 4.3). Large vertical cracks can also be observed on different zones of the cliff. As it was observed after the event, the geometry of the blocks remains strongly correlated to the initial fracture network of the rock mass.

Figure 4.2. *Location a) and shapes b) and c) of the blocks positioned on the slope. For a color version of this figure, see www.iste.co.uk/richefeu/gravity.zip*

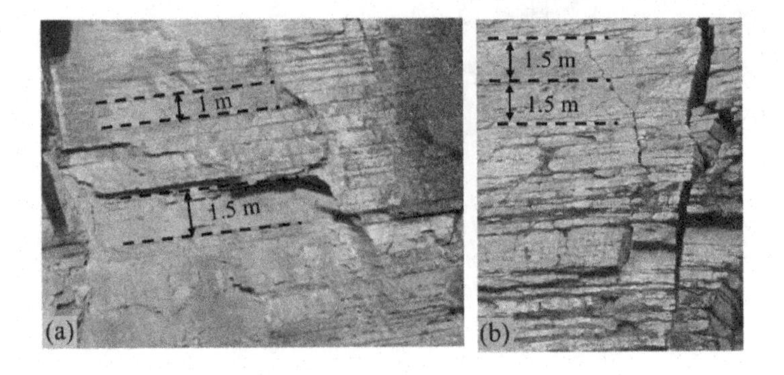

Figure 4.3. *Views of the horizontal a) and vertical b) fracture networks*

4.1.3. *Numerical model and parameters*

The issue addressed here is how an extreme simplification of the cliff surface and propagation area will affect the quality of the model. Two *digital surface models* (DSMs) were employed for this purpose. In the first DSM (Figure 4.4), the propagation slope and the cliff surface are simply represented by a succession of inclined planes. The second DSM (Figure 4.5) restores the actual topography of the site; it has been reconstructed by photogrammetry techniques. In both cases, the shape of the numerical blocks remains as faithful as possible to the actual geometry of the blocks identified in the field. In a first approach, only the large blocks were considered. For that, the unstable mass was cut by horizontal and vertical plans according to the fracture network identified on the cliff and as a function of the block sizes observed in the deposit area (Figure 4.1). With the simplified DSM, a total of 122 rock blocks, 1,050 m^3 in volume, were used (81 blocks ranging from 5 to 15 m^3 while 41 blocks sized less than 5 m^3). With the second DSM, 91 blocks were used, totaling a volume of 1,143 m^3 (one-third of small cubic blocks and two-thirds of large elongated parallelepiped blocks).

The contact parameters were derived from data available in the literature and adapted slightly as the result of a numerical sensitivity analysis. It is often said in the literature that the value of the normal dissipation coefficient e_n^2 ranges between 0.1 and 0.3 for boulder collisions on soft soil, and between 0.2 and 0.5 for rock collisions on granular layers. For this application, a value of 0.5 was retained to account for the soft nature of the slope soil (block/soil collisions) while a value of 0.45 was retained for the interactions between blocks. The friction angle between two blocks was measured by an elementary tilting test to a value of 45°, so that the friction parameter μ between blocks was set to 1. The interaction mechanisms involved in the collision of a rocky block with the slope are more complex than those between two blocks; they include dissipative phenomena such as friction and buttress. To account for these two mechanisms, a value of 1 was also retained for the tangential dissipation parameter μ for the block/soil interaction. High values of the normal and tangential stiffness at the contact k_n and k_t were retained to ensure narrow penetration between the blocks and the soil regardless of the mass and velocity of the block. The values of the most important parameters are summarized in Table 4.1 for the block/soil and block/block interactions.

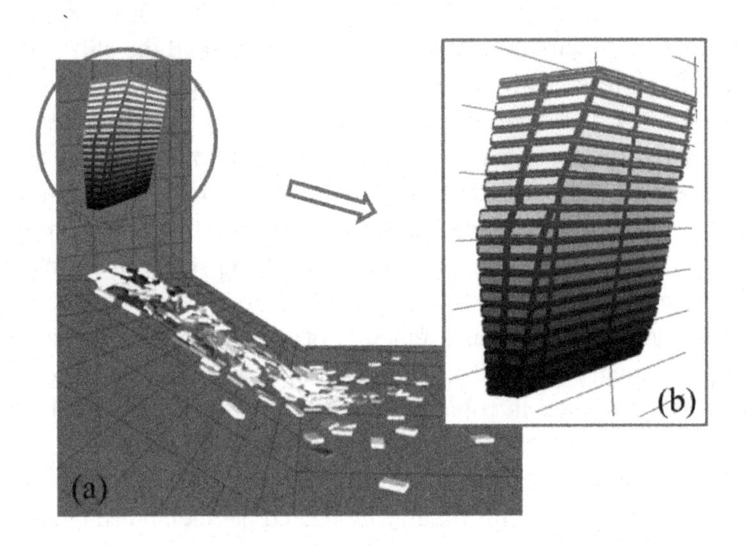

Figure 4.4. *Simplified digital surface model used to describe the natural terrain*

	e_n^2	μ	k_n (N/m)	k_t/k_n
Block/soil	0.15	1.0	10^8	1.0
Block/block	0.45	1.0	10^8	1.0

Table 4.1. *Numerical values of the contact parameters*

At the beginning, the blocks are destabilized by the gravity and fall down vertically near the cliff in a quasi-free fall movement that involves very few interactions with the cliff. Then, the rock mass hits the toe of the cliff and propagates along the slope. The blocks interact by collisions and friction between themselves and with the slope. Some isolated blocks are projected outside by their neighbors and propagate far away from the main deposit.

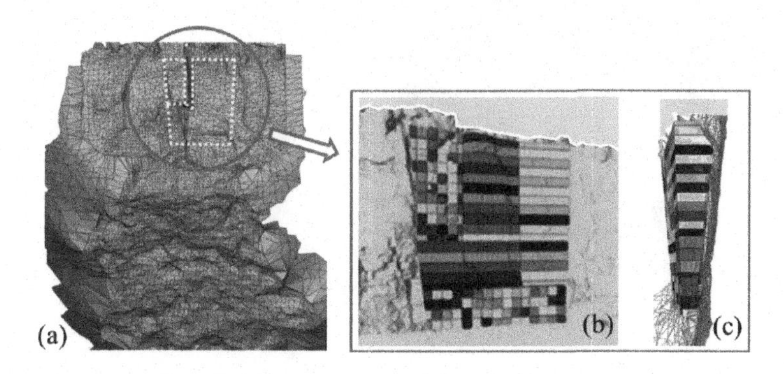

Figure 4.5. *Sophisticated digital surface model used to describe the natural terrain. For a color version of this figure, see www.iste.co.uk/richefeu/gravity.zip*

4.1.4. *Selected results*

4.1.4.1. *Use of a simplified DSM*

For this kind of event, the most important information for the characterization of the risk is the position of the blocks at the end of the rock avalanche. For this purpose, a comparison of the actual and simulated deposits is presented in Figure 4.6. We can see that most of the numerical blocks (96%) were stopped inside the same envelope as the real event. The 4% remaining blocks were found at the vicinity of the cliff toe. Other subtle behaviors were also restored by the simulation such as a high density of blocks concentrated at the base of the slope and notable positions of some isolated blocks (spotted in red in Figure 4.6). The main characteristics of the block deposit are summarized in Table 4.2 and Figure 4.7. Despite the simplistic nature of the DSM employed, a fairly good concordance of all the deposit characteristics is obtained.

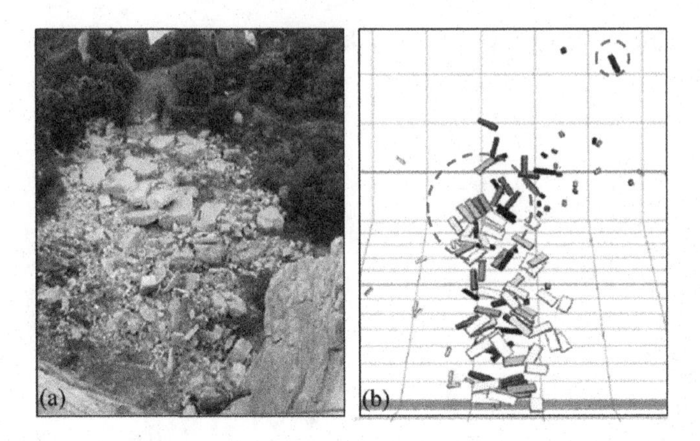

Figure 4.6. *Comparison of the actual deposit a) and the numerical deposit b). For a color version of this figure, see www.iste.co.uk/richefeu/gravity.zip*

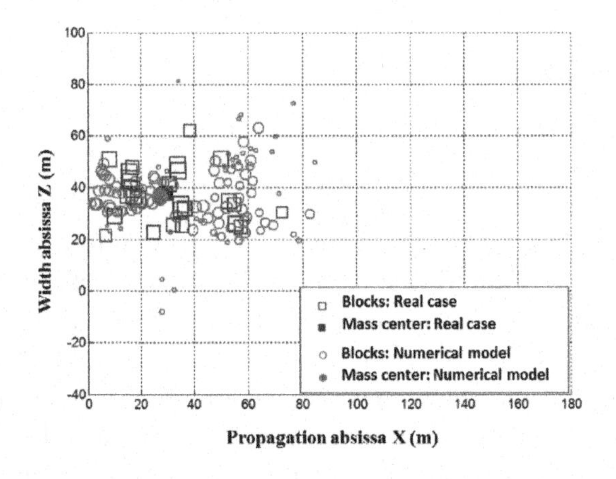

Figure 4.7. *Top view of the actual positions of the blocks superimposed with the positions obtained by simulation with the simplified DSM. For a color version of this figure, see www.iste.co.uk/richefeu/gravity.zip*

The energy dissipated by friction and collisions within the granular mass and at the base of the flow (hardly measurable *in situ*) is also followed during the mass motion. Figure 4.8 shows the location and intensity of the different modes of energy dissipation (block/block friction, W_{tBB}; block/soil friction, W_{tBS}; block/block collisions, W_{nBB} and block/soil collisions, W_{nBS}). Two different regions can easily be distinguished: (1) the first 10 m at the toe of the cliff where at least 41% of the total energy is dissipated, mainly by block interactions (36% by friction and 34% by collisions), and (2) the rest of the propagation path along the slope where the energy is mainly dissipated by friction between the blocks (25%) and by collisions of the blocks with the slope (55%). Interestingly, a sensitivity study (not reported here) has showed that for the tested geometry, the rate of energy dissipated at the cliff toe is always the same (about 40%) whatever the parameter used; it is thus not strongly related to the contact properties themselves.

	In situ measurement	DEM
Deposit width (m)	40	43.2
Deposit length (m)	66	68.5
Deposit thickness (m)	3	3.8
Runout distance (m)	67	70
Fahrböschung angle (°)	42	41
Travel angle (°)	45.5	47

Table 4.2. *Comparison between numerical and in situ characteristics of the deposit*

4.1.4.2. *Individual releases versus mass flow*

To get an idea of how the interaction between the blocks influences the propagation mechanisms, the deposits from the previous simulations have been compared with the final positions resulting from the release of individual blocks (Figure 4.9). The initial positions of the blocks in the cliff and their shapes were kept the same. Also, the numerical parameters were the reference ones (see Table 4.1). In Figure 4.9(a), the size of the markers refers to the initial block volume. The blocks are colored according to their initial position on the cliff: the blocks lying on the lower rows are colored in dark while the highest blocks are colored light brown. The lightly colored red lines correspond to the trajectory of each block. In this figure, we can see that the propagation distances are longer for individual releases than for collective flow. Approximately 20% of the blocks were stopped beyond the experimental front position and many blocks were scattered over the width. The interactions between the blocks when considering a collective flow enable some dissipation mechanisms that make the runout distances shorter. This is a clear benefit of the DEM in comparison to usual trajectory analysis (see section 1.1).

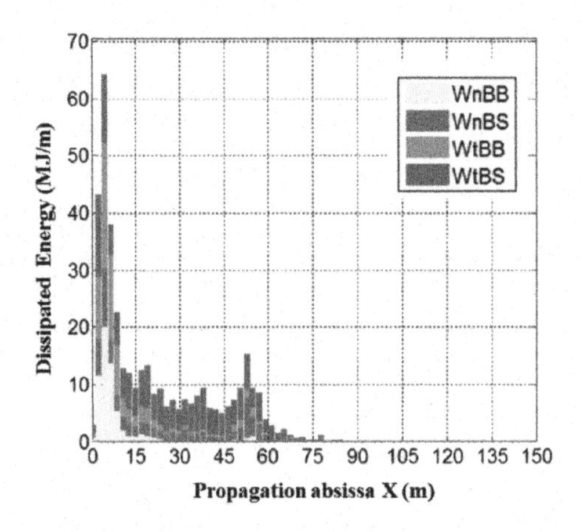

Figure 4.8. *Modes of energy dissipation along the propagation path. For a color version of this figure, see www.iste.co.uk/richefeu/gravity.zip*

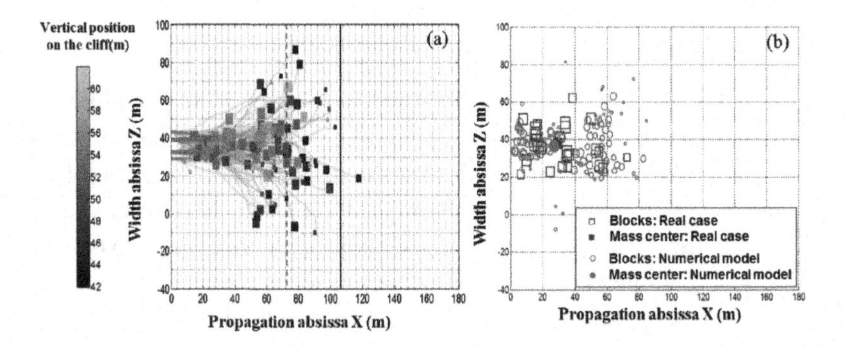

Figure 4.9. *Top views of the final positions of the blocks when released individually a) or all together b). For a color version of this figure, see www.iste.co.uk/richefeu/gravity.zip*

The translational and rotational velocities of the smaller and the bigger blocks that have the farthest runout distance are plotted in Figure 4.10 for both individual releases and collective flow. Note that the shape of the smaller block is rather cubic while the larger block in an elongated parallelepiped. We see that the translational velocities of the smaller block are of the same order in both cases (individual releases or collective flow). The larger block, because of multiple interactions within the mass, exhibits very different kinematics. The rotational velocities are lower in the mass movement, especially for elongated blocks such as the bigger one. Once again, the crucial role of the interactions within the flowing mass, which are not involved in trajectory analysis, is demonstrated.

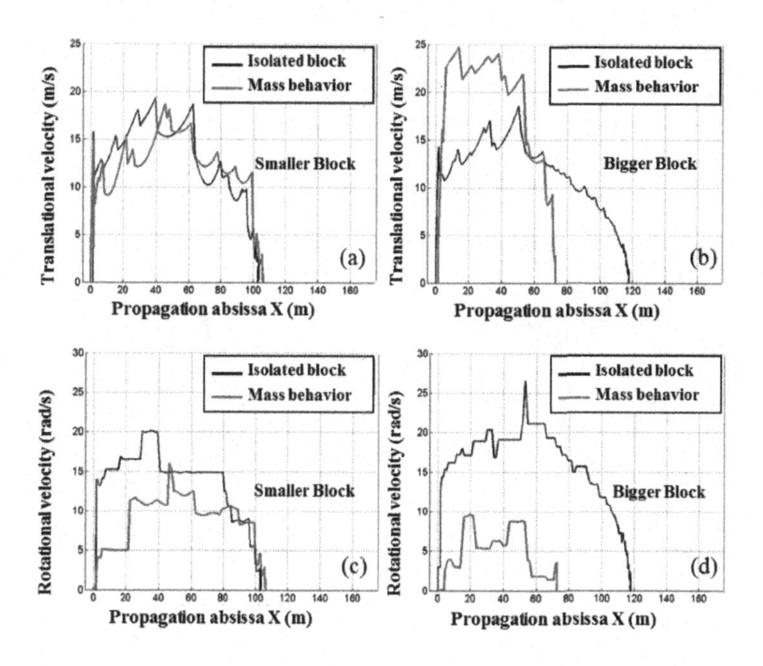

Figure 4.10. *Translational (a and b) and rotational (c and d) velocities for smaller block (a and c) and bigger block (b and d). For a color version of this figure, see www.iste.co.uk/richefeu/gravity.zip*

4.1.4.3. *A more realistic DSM*

A more realistic geometry of the propagation slope has been used. The DSM has been reconstructed from photographs shot by an unmanned aerial vehicle (a "drone"). The comparison with the previous simulation was achieved via the position of the final deposit and the amount of dissipated energies. Figures 4.11 and 4.12 show the different deposits of blocks: (a) *in situ* (b) with the simplified DSM and (c) with the realistic DSM. A better match is clearly observed with the scanned DSM. In particular, the lateral spreading and the main direction of propagation are better restored. The runout distance and the farthest isolated blocks are rather similar. We also note that, unlike the simplified configuration, the smallest blocks remain on the slope (due to surface roughness). Moreover, the large blocks stopped on the first part of the propagation area, which is consistent with the observations made on the site. Note also that the medium-sized blocks, located at the front of the rockfall, come from the upper part of the cliff (the lightest blocks were positioned in the upper layers of the stack while the darker blocks were positioned in the lower layers).

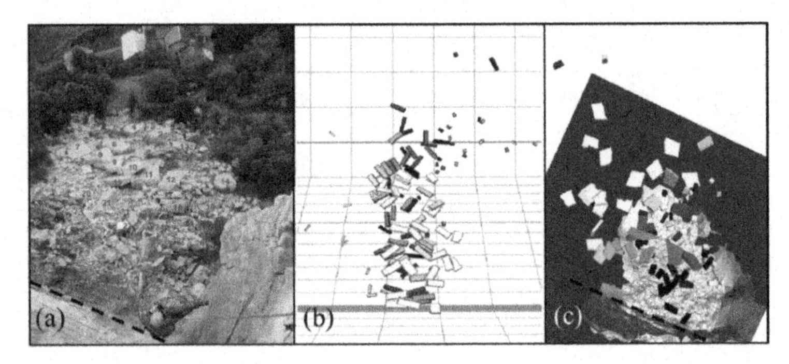

Figure 4.11. *A top view of the final deposits: a) in situ, b) with the simplified DSM and c) with a more realistic DSM. The colors of the blocks correspond to their initial position from bottom (dark) to top (light). For a color version of this figure, see www.iste.co.uk/richefeu/gravity.zip*

Figure 4.13 shows the time course of the dissipated energies accumulated by types. The roughness of the realistic DSM induces a perturbation in the granular flow that increases the dissipation within the granular mass (W_{nBB} and W_{tBB}). In contrast, the amount of energy dissipated by friction at the base of the flow is decreased because of the terrain roughness. Note that the same trend was observed in section 3.3 when analyzing the influence of the roughness of the slope on the propagation of bricks on planar or undulated slopes.

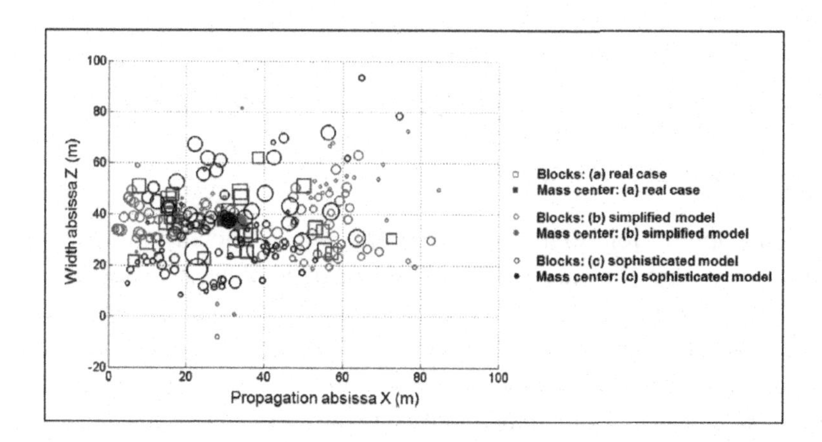

Figure 4.12. *Top view of actual and simulated final deposits. For a color version of this figure, see www.iste.co.uk/richefeu/gravity.zip*

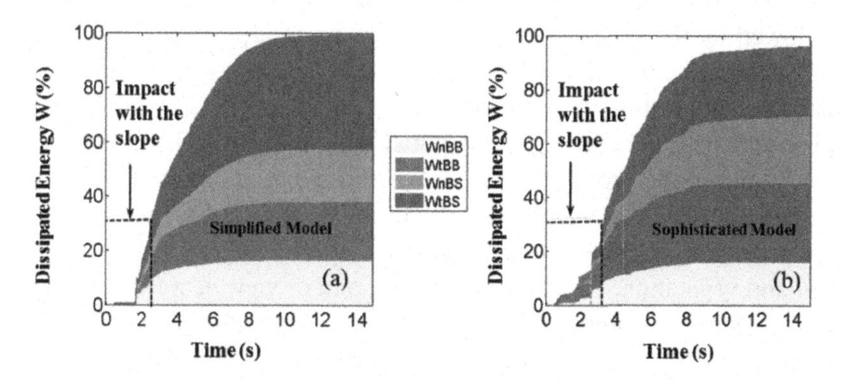

Figure 4.13. *Comparison between the dissipated energy modes between simplified a) and more realistic b) models*

To conclude, the comparison between the geometries of simulated and actual deposits showed that the numerical model is able to satisfactorily restore the kinematics of this type of rockfall provided that adequate contact parameters, realistic shapes of the blocks and the terrain are taken into consideration. Not only the runout distance, but also more subtle features such as the apparent shape of the deposit or the block arrangements (stacked or dispersed) were well restored by the simulations.

4.2. Numerical modeling of an artificially triggered rockfall

4.2.1. *Purpose of the modeling*

This application refers to an event that was artificially triggered in order to ensure the safety of neighboring habitations and population [BOT 14]. It was a blast of an unstable rock mass of approximately $2,600$ m^3. This scheduled event allowed scientists and stakeholders to make specific measurements before, during and after the blast of the rock mass. In particular, photogrammetry and aerial LIDAR techniques were employed before and after the event, allowing the characterization of the unstable volume and the fracture network. Photographs and videos of the event have provided the estimation of the velocity of the rock mass and several individual blocks. The rockfall event consisted of a free fall down to the toe of the cliff followed by the propagation of the blocks up to a protective structure (a ditch and an earthen barrier) far away the blasting zone. The particularity of this site is the very long distance covered by the blocks that separate quite early from the initial rock mass, to roll, bounce and spread along the propagation path. For this event, it was necessary to distinguish the mechanical parameters involved in different zones of the propagation path since they are made up of a very large variety of soil and vegetation. From the numerical point of view, it was necessary to introduce the rolling resistance mechanism because of the softness of the soil at certain zones (in the sense that the rock indentations were deep). The initial velocity of the blocks due to the blast has been found to be of primary importance and was estimated from the videos of the event.

4.2.2. *Description of the site*

The eastern limestone cliff of the mountain called Néron (located at 5 km NW of Grenoble, French Alps, Isére, France) was the theater of two medium-sized gravity events during summer and winter 2011. On August 14, 2011, a 2,000 m^3 rock compartment detached from the cliff, fell 100 m below and propagated down the slope. Although most of the fallen rocks deposited in the upper part of the slope, several 15 m size blocks were stopped by a ditch and an earthen barrier after a propagation distance of 800 m. An unstable overhanging 2,600 m^3 compartment remained attached to the cliff and was blasted on December 13, 2011. During this artificially triggered event, seven blocks reached the ditch, with volumes ranging from 0.8 to 12 m^3.

Figure 4.14 shows an image of the Néron site after blasting. The upper part of the site is a quasi-vertical cliff of 150 m high. Below the cliff, there is a 40° inclined slope of approximately 850 m long. The first part of the slope is covered by scree deposits overlying marly limestone while the second lower part shows gentle slope covered by trees and soft marl outcrops. At the toe of the slope, a ditch was dug, and a 9 m high by 300 m long earthen barrier was built to protect the neighboring population.

The dangerous unstable mass was located on the top part of the cliff (marked (1) in Figure 4.14). During the first event, 2,000 m^3 of the lower part of the unstable column (in blue on Figure 4.15) detached from the cliff, the 2,700 m^3 upper compartment of the rock column (in red on Figure 4.15) remained attached to the cliff until the decision was taken to blast it by means of 400 kg of explosives.

From the video shots, several phases of the rockfall could be identified. The mass that was blasted, fell freely during 3–6 s before reaching the toe of the cliff (Figure 4.16); it then propagated down the scree slope by generating a dense aerosol. Most of the fallen volume was deposited in the upper scree area while only a few large blocks propagated downwards across the forest, they bounced on the ground and cut down trees (Figure 4.16). Seven-meter size blocks, circled in red and numbered from 1 to 7 in Figure 4.17, with volumes of 3, 0.8, 4,

12, 9, 9, and 4 m, respectively, reached the toe of the slope and were stopped by the ditch-earthen barrier system about 45 s after the blast.

Figure 4.14. *A photograph of the Néron site: (1) the rockfall scar (light gray), (2) the zone of mass inlet in the lower part of the cliff, (3) the scree deposits, (4) the forest and (5) the ditch and earthen barrier structure*

The velocities of the blocks after blasting and at the early stage of the fall were estimated by means of the Digital Image Correlation technique on video shots acquired from a fixed camera. No 3D reconstruction has been used so that the velocities considered hereafter are actually those projected on the photograph plan. However, these data are still exploitable because of the long distance of the blasting, making the vertical component of the measured displacements very slightly distorted. Also, since the cliff is roughly parallel to the camera plane, the out-of-plane displacements were lower by several orders of magnitude. The block velocities are spatially averaged since one pixel may "hold" several blocks. The displacement field (magnified 10 times in pixel size) at three different time points (1, 3 and 11 s) are superimposed with the corresponding images in Figure 4.18. From these data, it was deduced that the initial horizontal velocities of the blocks due to the blast ranged between 10 and 15 m/s. At the end of the free fall, the maximal block velocity was estimated at approximately 30 m/s, but from the toe of the cliff, the fallen volume behaved as a rock mass with an initially moderate propagation velocity of 20 m/s that settled rapidly.

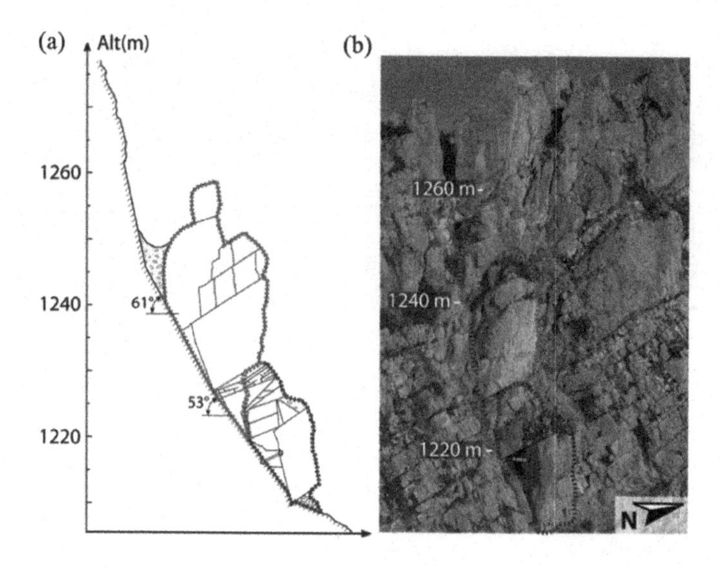

Figure 4.15. *Details of the two rock compartments on the cliff: a) initial detached mass in blue; b) blasted mass in red. For a color version of this figure, see www.iste.co.uk/richefeu/gravity.zip*

Figure 4.16. *Video shots of the event: blasting a), a bouncing block b) and rolling block c)*

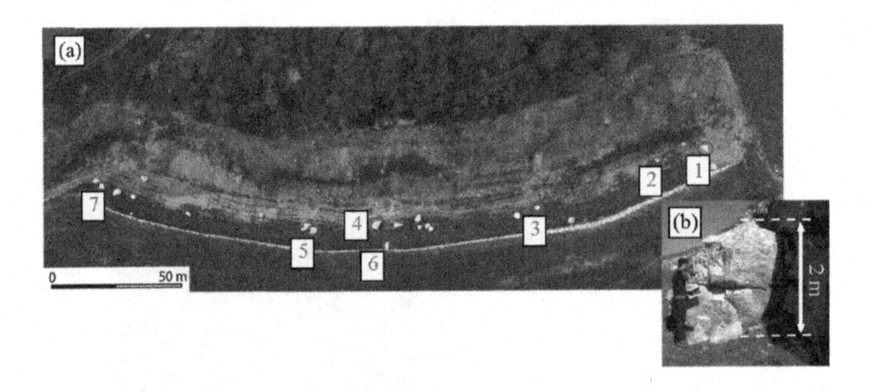

Figure 4.17. *Location of the blocks that stopped at the ditch-earthen barrier*

The translational and rotational velocities of an isolated block (numbered 6) propagating down the slope were determined by the use of stereographic videos of the blasting. Two specific bounces occurring at mid-slope and showing long free-flight phases were studied in detail. The block translational velocities before and after the two bounces ranged from 12 to 14.5 m/s and from 22 to 28 m/s, respectively. At the vicinity of the ditch-earthen barrier, the velocities of block number 6 were estimated at approximately 20–25 m/s in translation and approximately 10 rad/s in rotation.

Figure 4.18. *Displacement fields just after the blast. For a color version of this figure, see www.iste.co.uk/richefeu/gravity.zip*

4.2.3. *Numerical model and parameters*

The triggering of the rockfall consisted of assigning an initial horizontal velocity (V_0) to the blocks of the unstable rock mass to emulate the initial blasting energy. The value of V_0 was set at 10 m/s outwards and normal to the cliff, which is the value suggested by the digital image correlations. In this particular instance of blasting, the shape and the size of the blocks are caused by the mining network rather than by the natural fracture network. As a result, the unstable rock mass – defined by subtracting the DSMs before and after the event – is modeled by randomly arranged blocks with an apparent volume of about 2,500 m^3. Each rock block is a regular sphero-polyhedron with a unitary volume uniformly distributed between 1 and 9 m^3, matching the size of blocks observed during the rock propagation along the inclined slope in the real event. Smaller blocks probably played a certain role in the early mass flow stage of the rockfall but then settled rapidly. For this reason and for saving computation time, they were not included in the simulations.

The DSMs for Néron's site were built from aerial LIDAR measurements conducted both before and after the triggered event. The points corresponding to the ground and cliff surface were kept by filtering those with a low reflectivity coefficient (reflections on the vegetation) and by manually rejecting aberrant points. The numerical topography of the slope retained for the numerical modeling corresponds to a coarse version of the LIDAR DSM (subsampled down to triangular elements of 10 m in characteristic size) to speed up the computation. The propagation area was zoned in four different parts according to the type of terrain: solid rock (1), rocky scree (2), soft ground with trees (3) and earthen barrier material (4) (Figure 4.19). For each type of contact, the dissipation coefficients along the normal direction (e_n^2), the tangential direction μ and the rolling friction (μ_R) were chosen from a set of values derived from the literature and improved by trial and error until the simulated time propagation behaved closely to the real one. The values of the main numerical contact parameters are summarized in Table 4.3. Notice that the elasticity involved in the rolling resistance law was replaced by a viscous parameter γ_R so that the resistant moment $\mathbf{C}$ was proportional to $\gamma_R \mathbf{\Omega}$, where $\mathbf{\Omega}$ is the angular velocity of the block acting on the soil. This viscosity was, however, not so crucial and could have been replaced by an angular elasticity.

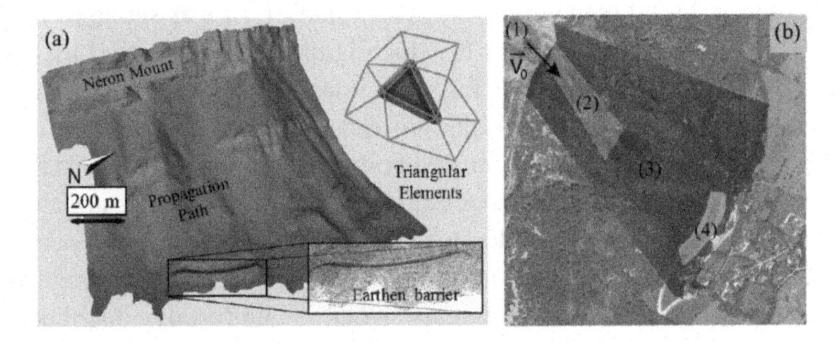

Figure 4.19. *Digital surface model and spatial zoning of the site*

4.2.4. *Selected results*

Immediately after the triggering, consisting of the setting of an initial velocity of the blocks of 10 m/s, the blocks remained together as a whole and fell freely down to the toe of the cliff. Similarly to the real event, most of the blocks (nearly 80%) stopped when reaching the rocky scree at the top of the slope because of rolling resistance. The total duration of the event was about 80 s, with the first arrival at the earthen barrier occurring at $t = 40$ s, which is similar to the real event. Because of the DSM resolution and maybe a lack of relevancy in some dissipation parameters – that are actually difficult to assess – the positions of the stopped block in the simulation could not be directly compared with the experimental ones. However, 90% of the block paths were in good accordance with the propagation area of the real event. The blocks that did not deposit on the rocky scree (20% of the total mass) propagated down the slope at a mean velocity slightly under 10 m/s, with very rare collisions occuring between them. Six blocks reached the ditch-earthen barrier with a velocity of 15–25 m/s just before the impact. The kinematics of the numerical flow is shown in Figure 4.20 by means of temporally ordered snapshots of the simulation. We can observe that the global event is well reproduced, showing the ability of the numerical model to chart this kind of phenomenon.

	e_n^2	μ	μ_R	k_n (N/m)	k_t/k_n	$\gamma_R(N.m.s)$
Block/rock	0.3	0.5	0.0	10^8	1.0	10^7
Block/rocky scree	0.05	1.0	0.2	10^8	1.0	10^7
Block/soft ground	0.01	0.2	0.1	10^8	1.0	10^7
Block/earthen barrier	0.001	1.2	0.9	10^8	1.0	10^7
Block/block	0.3	0.5	0.0	10^8	1.0	10^7

Table 4.3. *Values of the contact parameters for the simulation of Néron's event*

Figure 4.21 shows the mean vertical velocity V_z and the mean velocity magnitude V_n of the blocks at the very beginning of the event (first 17 s). These quantities are obtained (1) from the simulation, (2) from the captured image analysis and (*iii*) from the analytic solution of a free fall:

$$V_z = gt \quad \text{and} \quad V_n = \sqrt{(gt)^2 + V_0^2} \qquad [4.1]$$

where $g \simeq -9.81$ m/s^2 is the gravitational acceleration. All curves display a rapid increase in both velocities during the first 3 s, following the theoretical free-fall equations. The 30–35 m/s maximum mean block velocity is reached 4–5 s after the blast, before a rapid slow down (from 5 to 6.5 s), associated with the block inlet on the ground. Then, the velocity decreases smoothly between 6.5 and 17 s, as several rock blocks slow down and deposit. The overall shape of the numerical time–velocity curves is in good accordance with the data from the digital image correlations, with consistent maximal mean block velocity and free-fall duration. Average numerical velocities are lower for times greater than 7 s, because of the restricted observation window when processing the digital images and because the small blocks continuously falling down the cliff were not taken into account in the numerical model.

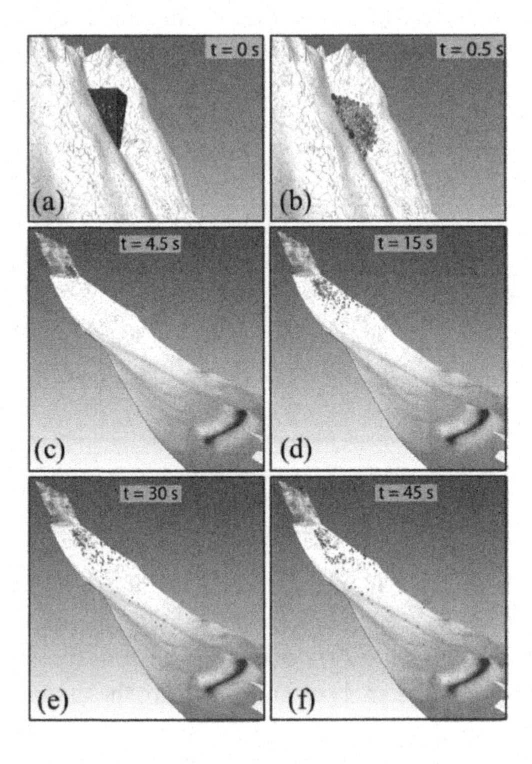

Figure 4.20. *Snapshots of the simulation at different times. For a color version of this figure, see www.iste.co.uk/richefeu/gravity.zip*

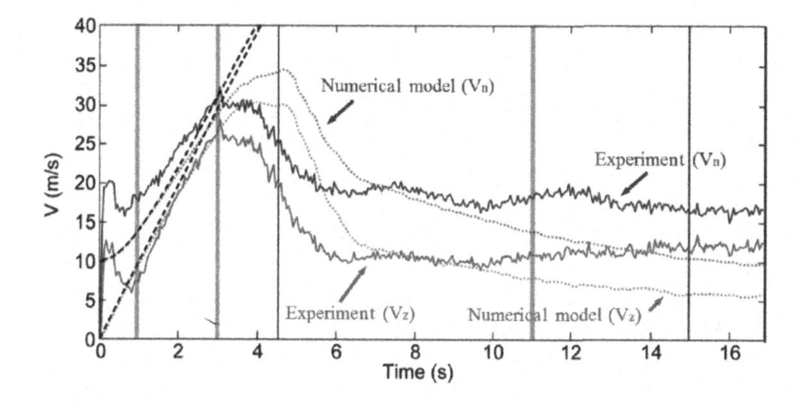

Figure 4.21. *Comparison between the mean vertical velocity V_z and the mean magnitude of velocities V_n of blocks after blasting, obtained from the simulation (dashed color lines), estimated from the video shots (solid color lines) and resulted from analytic derivation of a free fall (dashed black lines). For a color version of this figure, see www.iste.co.uk/richefeu/gravity.zip*

Complementary numerical simulations were performed to establish the main parameters controlling the rockfall kinematics. First, it was shown that the rockfall propagation is highly sensitive to the blast modeling: the initial velocity orientation strongly controls the propagation path, whereas its magnitude is less crucial provided that it remains in the 5–20 m/s range. Second, the resolution of the DSM plays a major role in the block propagation, especially in zones receiving numerous collisions and located uphill. The selection of an appropriate resolution could be addressed by means of sensitivity studies or by comparison with well-documented rockfall events. In our case, the 10 m typical size for the DSM appeared too coarse for meshing the toe of cliff area, where most of the rock blocks collided after free-fall. As this zone lies at the top of the slope, it strongly affected the deposit pattern. Finer DSM should significantly improve the results, but would require more computation time. Moreover, the propagation path of isolated blocks and the arrival times of the first and last blocks in the ditch-earthen barrier were consistently reproduced. The model-predicted velocities were comparable to those estimated from the videos, i.e. ranging from 20–30 m/s. The key points to obtain realistic numerical simulations are as follows: a sufficiently high resolution of the digital surface, appropriate initial velocities (for a triggered event) and the use of appropriate dissipation coefficients (especially the rolling resistance imposed by the soft ground).

4.3. Forecast of a rockfall propagation toward a protective structure

4.3.1. *Purpose of the modeling*

This application deals with the case of an unstable rock mass that could separate from a vertical cliff and hit an existing protective wall [CUE 15]. In this case, we were interested in observing the ability of the model, firstly, to simulate the vertical free fall of the blocks and the motion of the block mass after the first collision, and secondly, to define the amounts of energy that would be lost or transmitted to the protective structure. The discrete model makes it possible to obtain the location and the intensity of the block collisions on the protective wall as well as the collective motions and interactions of the blocks in the falling mass.

The contact parameters were deduced from experimental tests, and the geometries of the cliff and of the unstable mass were obtained from photogrammetric analysis. Since the granular mass principally fell down and stopped at the toe of the cliff, it was not necessary to introduce any rolling resistance between the soil and the blocks. In a simulation, the energy is thus dissipated only by collisions and friction, whereas within the granular mass or for interactions of the blocks with the soil and the wall.

4.3.2. *Description of the natural site*

The unstable rock mass, which had not yet fallen from the cliff when writing this book, represented a risk for the bordering zones. The site is located in the Vercors Mountains in the French Alps, close to the city of Grenoble. It consists of an almost vertical rocky cliff of 60 m height constituted of Paleocene limestone layers of approximately 0.3 m thickness (Figure 4.22), which exhibits some signs of instability such as small rockfalls. Some residential buildings and roads are located next to the cliff, which increases the risk in this area (Figure 4.23).

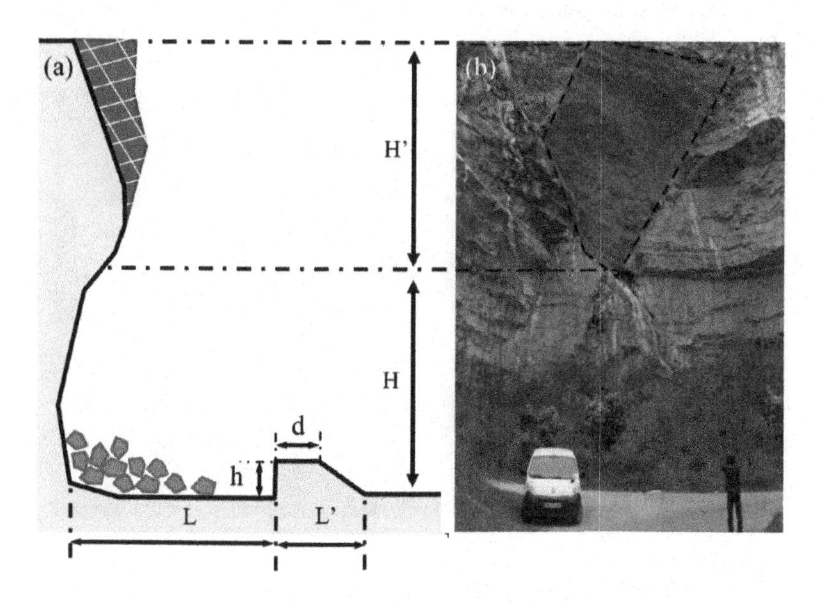

Figure 4.22. *Geometry of the cliff a) and dimensions of the unstable zone b)*

The protective structure was built in 2007, to protect the nearest residents from minor rockfall events smaller than 200 m^3 in volume. The protective wall is located approximately 30 m from the cliff. The average dimensions of the protective structure are as follows: 8.4 m width at the base, 3 m width at the top, a vertical front face of 3 m height. It was observed that the blocks that had already fallen far enough from the protective structure and did not represent a risk to people and the structure itself. The size of the biggest blocks ranged between 1–2 m (Figure 4.24).

Some significant displacements were recently recorded at the top of the cliff, suggesting a possible large rockfall, making it necessary to perform a redesign study of the structure. A geological survey was carried out in 2013 and a potentially unstable rock mass was notably observed. It consists of a 20 m high and 40 m wide area limited by two fracture planes, oriented at 1° N/66° and 105° N/76°, respectively. As a result, the potential unstable volume was estimated to be approximately 2,500 m^3.

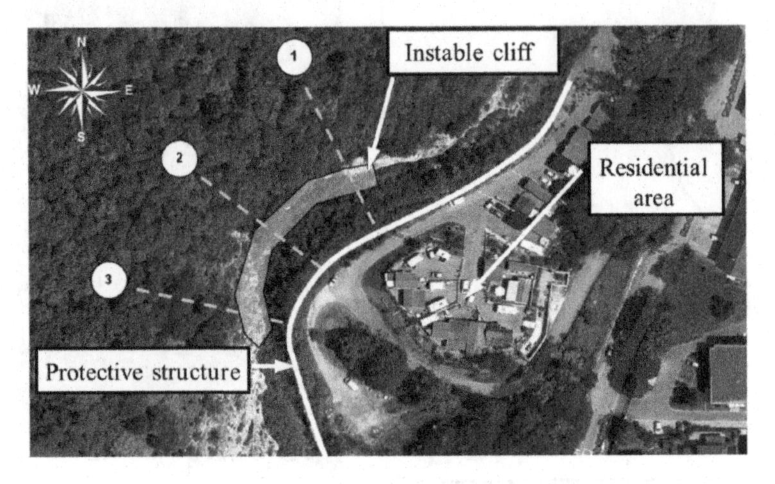

Figure 4.23. *Aerial view of the cliff, protective structure and residential area*

Figure 4.24. *Size of the fallen blocks at the toe of the cliff*

4.3.3. *Numerical model and parameters*

From the 3D reconstruction of the cliff relief by means of photogrammetry, three discontinuity families were pointed out on the front face of the cliff (Figure 4.25). By extrapolation, the unstable volume has been fabricated. This volume was subsequently subdivided into parallelepiped elements irregularly arranged along orthogonal planes. Given the uncertainties on the possible size of the blocks during the collapse of the unstable part of the cliff, two block sizes were considered for comparison purposes: 338 blocks for which the size ranges between 1.5 – 2 m (large blocks) and about 1,890 blocks, in which the size varies between 1 – 1.5 m (small blocks). In order to get a more precise idea of the actual ability of the structure to contain the detached mass, numerical simulations using 338 blocks and without protective structure were also performed.

The contact parameters are deduced from *in situ* observations or derived from values of the literature. The internal friction angle of the rock mass was estimated from the slope angle of old deposits and taken as the Coulomb friction angle between the blocks. The dissipation energy between blocks and the protective structure was assumed to be greater than the energy dissipated by collisions within the rock mass. The values of the main numerical parameters are summarized in Table 4.4.

	e_n^2	μ	k_n (N/m)	k_t/k_n
Block/natural ground	0.1	1.1	2×10^{10}	1.0
Block/block	0.1	1.1	2×10^{10}	1.0
Block/protective wall	0.01	1.1	2×10^{10}	1.0

Table 4.4. *Values of the contact parameters*

4.3.4. *Selected results*

It is with no surprise that in the simulations, the unstable rock mass underwent a free fall and were stacked at the toe of the cliff. It was, however, observed as an energy transfer that propelled some blocks beyond the protective structure. Figure 4.26 shows the comparison between the granular mass deposits for the three configurations (large blocks, small blocks and in the absence of protective wall). We can see that the overall shapes of the deposits are fairly similar; the essential differences between the two configurations with the wall are mainly related to the number of blocks that crossed the wall. Ultimately, 8 of the 338 blocks passed above the protective structure when using large blocks of 2 m, whereas only 3 of the 1,889 blocks passed the wall for small blocks of 1 m. In both configurations, the lateral spreading of the main deposit was approximately 71 m. In any case, the capacity of the protection structure was insufficient to retain all blocks.

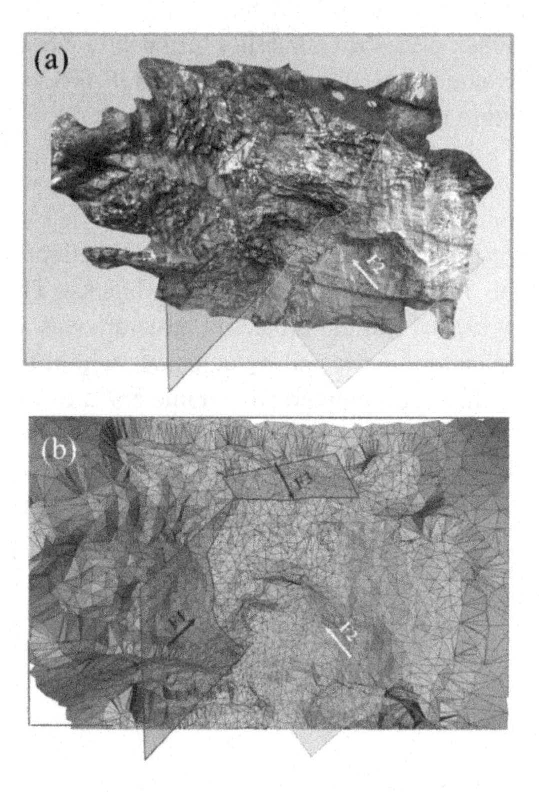

Figure 4.25. *3D reconstruction of the cliff, with the main orientations of discontinuities that have been identified from an analysis of the front face of the cliff. For a color version of this figure, see www.iste.co.uk/richefeu/gravity.zip*

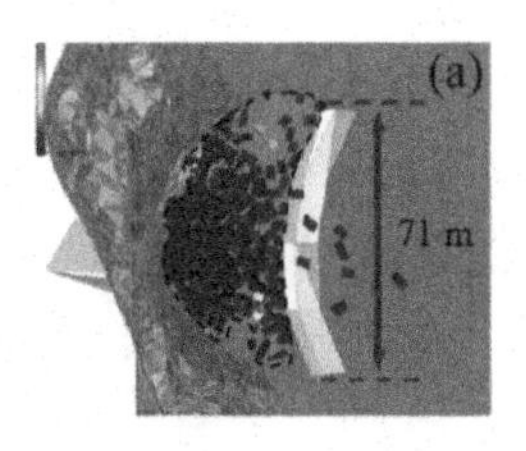

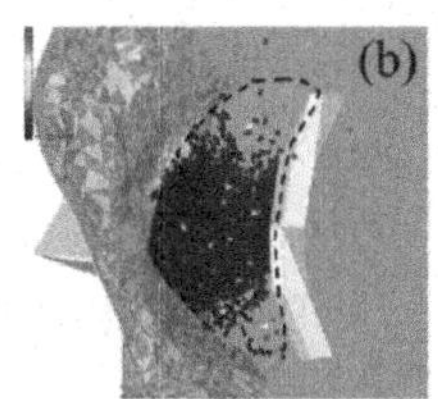

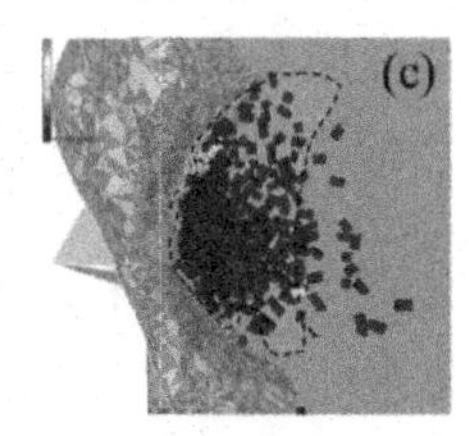

Figure 4.26. *Deposits obtained with large blocks a), small blocks b) and without protective structure. For a color version of this figure, see www.iste.co.uk/richefeu/gravity.zip*

Two phenomena can be assumed to explain the ability of large blocks to cross the protective structure, both linked to the block sizes. The first is related to the overall volume of the deposit, which is increased when the blocks are oversized, due to the large voids induced by boundary effects with the natural terrain and the wall itself. This increase in the height of the deposit facilitates the overflow of the blocks above the protective wall. The second assumption is that the increase in the number of interactions for small blocks smooth the whole mass motion, so that all block trajectories are quite undisturbed (i.e. with lower energy of ejection). On the contrary, large blocks can be propelled far away because of a massive transfer of energy from one block to another. Furthermore, as illustrated in Figure 4.27, the motion of large blocks reflects an important mass translation further from the protective structure, while the small blocks have higher rotational velocities. The orders of magnitude of the translational velocities are about 10–20 m/s while the rotational velocities reach a maximum of about 10 rad/s for the small blocks. The spatial distribution of velocities on the protective structure is shown in Figure 4.28 for large and small blocks. It may be noted that the collision velocities are fairly similar in both cases, but they are more uniformly distributed when the blocks are small. In terms of collision energy, that is when considering together velocities and masses, the total amount of energy transmitted to the protection structure is quite similar.

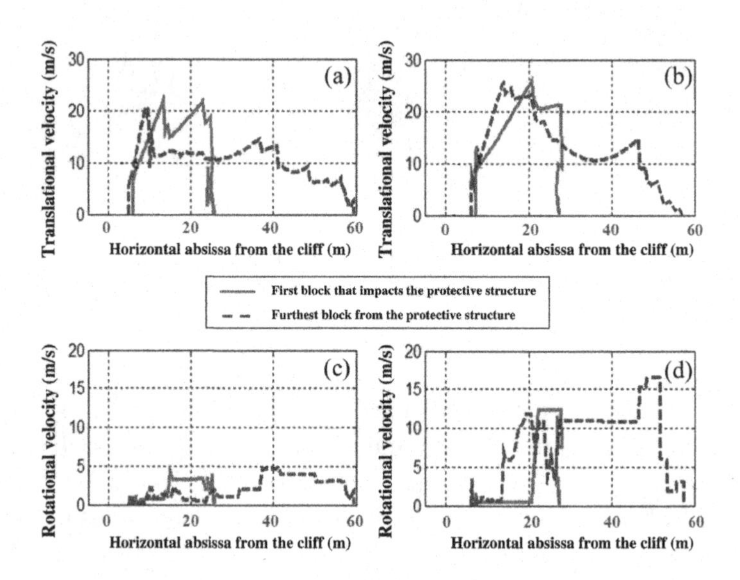

Figure 4.27. *Translational (a and b) and rotational (c and d) velocities of the blocks for the numerical configuration using large blocks (a and c) and small blocks (b and d)*

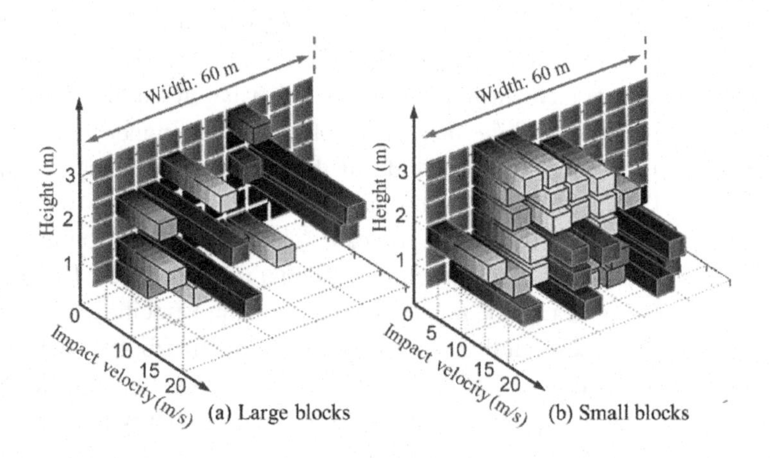

Figure 4.28. *Collision velocities on the protective structure large blocks a) and small blocks b)*

For the simulation performed without the protective structure, shown in Figure 4.26(c), we obtained, as expected, a more widespread deposit with a number of isolated blocks that traveled a longer distance when compared with the configuration protected by a wall (8 blocks crossed the wall while more than 20 blocks passed through the same limit in a configuration without any protective structure). From these results, it was decided to increase the height of the protective structure to minimize the number of isolated blocks crossing the protective wall.

5

From Discrete to Continuum Modeling

In this chapter, two numerical methods – DEM and MPM – are compared on the basis of numerical simulations, which consists of the release and propagation of a granular material in an inclined channel ending with a horizontal stop area (Figure 5.1). Randomly packed blocks (cubes of $0.01 \times 0.01 \times 0.01$ m^3 or cuboid elements of $0.02 \times 0.01 \times 0.005$ m^3) were used for DEM simulations while a continuum mass governed by an elastoplastic law was considered for MPM simulations. The release container was positioned on a 45° slope at a given height. Between the slope and the horizontal stop area, a smooth transition was set by means of a curvature of 0.3 m in radius. DEM simulations were performed in a three-dimensional configuration; the flow was thus canalized to make the comparison possible with the two-dimensional MPM simulations. The channel width was set to 0.2 m to make the flowing truly three-dimensional while limiting the number of elements.

The work presented in this chapter is the result of collaboration with Fabio Gracia, of the IMSRN company (www.imsrn.com).

5.1. Geometries and parameters used

To provide some variability, we used different container sizes, basal friction coefficients, height of release and shape of blocks (only for the DEM). Different aspect ratios of the container (parallelepiped boxes of

$0.246 \times 0.2 \times 0.2$ m^3 and $0.43 \times 0.2 \times 0.1$ m^3 in volume) were tested for different release heights (h = 0.4 m, 0.8 m and 1.6 m) in order to vary the total potential energy. Note that, depending on the container shape, the volumes of material, and consequently the initial potential energy, are not the same.

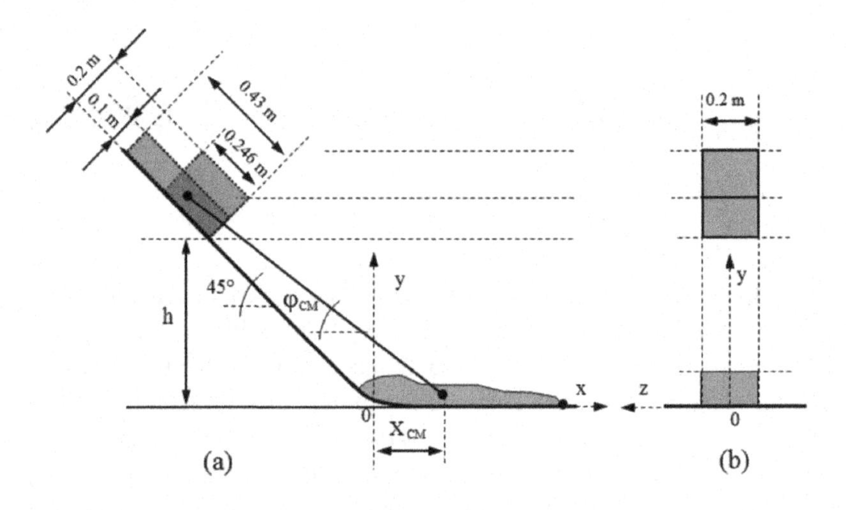

Figure 5.1. *Simulation setup: a) lateral view and b) front view showing the channel width for 3D-DEM simulations*

For the MPM model, a Mohr–Coulomb constitutive model was employed. The complete eslastoplastic model was chosen as it was the simplest possible one, with only the strictly necessary features that are the classical Mohr–Coulomb yield surface together with a non-associated flow rule. The parameters of the elastoplastic model are given in Table 5.1. The Eulerian structured grid was constituted with squared quad elements of 0.015 m $\times$ 0.015 m. Each material point had an initial volume of $0.005 \times 0.005 \times 0.2$ m^3.

The contact parameters governing the mechanical behavior of the DEM elements are given in Table 5.2. Two vertical frictionless walls were used at each lateral side of the channel to limit their influence on the flow.

Parameter	Notation	Value
Young's modulus	E	80×10^4 Pa
Poisson coefficient	ν	0.42
Internal friction angle	φ	28.6°
Coulomb cohesion	c	~ 0 Pa
Dilatancy angle	ψ	$\sim 0°$

Table 5.1. *MPM internal parameters*

k_{nBB}	k_{tBB}/k_{nBB}	μ_{BB}	e^2_{nBB}
10^4 N/m	0.4	0.7	0.2

Table 5.2. *DEM contact parameters*

In both cases (MPM and DEM), the interaction at the base of the flow is governed by the same contact law taking account of dissipation by collisions and friction. Two values of the basal friction angle, lower than the internal friction angle, were considered (16.7° and 26.6° corresponding, respectively, to the friction coefficients 0.3 and 0.5) in the study. The contact parameters are presented in Table 5.3.

k_{nBS}	k_{tBS}/k_{nBS}	e^2_{nBS}
10^4 N/m	0.4	0.2

Table 5.3. *DEM and MPM contact parameters involved at the base between the mass (or the blocks) and the support*

Nine numerical configurations were defined by varying the height of release, the shape of the blocks, and the shape of the container. These configurations are summarized in Table 5.4. Obviously, the shape of the blocks was not taken into account in the MPM simulations. For each configuration, the basal friction coefficient μ_{BS} was also varied. In order to be able to compare the three-dimensional DEM simulations and the two-dimensional MPM simulations, an equivalent mass density was used for MPM simulations.

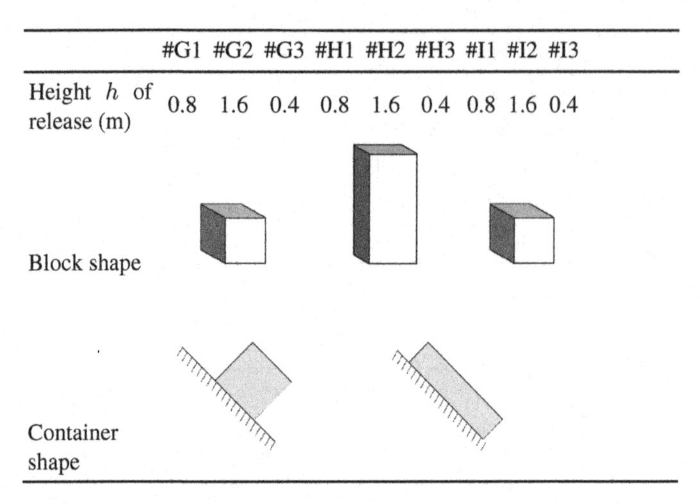

	#G1	#G2	#G3	#H1	#H2	#H3	#I1	#I2	#I3
Height h of release (m)	0.8	1.6	0.4	0.8	1.6	0.4	0.8	1.6	0.4
Block shape									
Container shape									

Table 5.4. *Configurations defined by changing the release height, the shape of the blocks and the shape of the container*

5.2. Analysis

Comparisons between the two models were first based on the kinematics of the flow. The discrete and continuous deposits were also compared in terms of position of the center of mass X_{CM} (called *propagation distance* here) together with the distribution of the masses around it. The latter was quantified by the standard deviation σ_{X} of the positions along the propagation path of all blocks or material points depending on the method. For both methods, the overall cumulated dissipation was split into three contributions: (1) the cumulated work W_n of the normal forces (contacts and collisions) at the base of the flow, (2) the cumulated work W_t of the tangential forces (friction) at the base of the flow and (3) the work W_{bulk} of all internal action in the flowing mass. We recall here that the interactions of the mass with the support is exactly the same for the DEM and MPM. The computation of the dissipation in the bulk is different for the two methods. For the MPM, the dissipation within the mass is:

$$W_{\text{bulk}}(t) = \sum_{p} \int_{0}^{t} (V_p \boldsymbol{\sigma} : \dot{\boldsymbol{\varepsilon}})\, \mathrm{d}t \qquad [5.1]$$

where V_p, σ and $\Delta\varepsilon$ are, respectively, the volume, stress, and strain increment held by each material point p. For the DEM, the dissipation within the mass is simply obtained as the sum of cumulated works of contact forces between the blocks in the normal and tangential directions

$$W_{\text{bulk}}(t) = W_{nBB}(t) + W_{tBB}(t) \qquad [5.2]$$

5.2.1. *Relatively low friction at the base*

All configurations were first tested with a relatively low friction coefficient at the base: the angle of friction was set to 16.7°, which is right below the internal angle of friction of 28.6°. The results obtained in terms of propagation and spreading of the mass, and total dissipation modes are summarized in Table 5.5 for both MPM and DEM simulations. We can already note a good match of the MPM and DEM simulations on all items. Let us now discuss them in more detail.

Figure 5.2 shows the kinematics of the avalanche by means of time shots of the lateral view of the flowing mass for the configurations #G3 and #I1. The blue shots are those of DEM and the red ones correspond to MPM simulations at the same moments. We see that the flow evolution from the beginning through to the very end correlates properly. Some minor discrepancies can be seen in the shape of the final deposit, as the MPM final deposit spreads a little further. The few blocks escaped from the mass that can be spotted at the front of the DEM deposit are completely absent in the MPM deposit for obvious reason.

In terms of center of mass (see Table 5.5), the value reached by both methods is very close in most of the simulation runs. This first observation is rather surprising when one realizes that the constitutive model used in the MPM is extremely simplistic. Indeed, we will see in the following that most of the dissipation arises at the base of the flow where only the contact model is involved – just like for DEM. The constitutive model is only involved in the stretching of the continuous mass.

▶ **MPM**, $\mu_{BS} = 0.3$									
	#G1	#G2	#G3	#H1	#H2	#H3	#I1	#I2	#I3
X_{CM} (m)	1.32	2.48	0.74	1.42	2.57	0.84	1.40	2.56	0.82
σ_{X} (m)	0.44	0.55	0.37	0.44	0.57	0.35	0.42	0.55	0.33
W_n (J)	0.15	0.38	0.10	0.09	0.19	0.05	0.09	0.20	0.05
W_t (J)	151.89	293.45	80.17	111.23	205.70	63.99	112.55	209.38	64.08
W_{bulk} (J)	15.49	18.12	14.65	6.70	8.80	5.37	6.65	8.76	5.39
▶ **DEM**, $\mu_{BS} = 0.3$									
	#G1	#G2	#G3	#H1	#H2	#H3	#I1	#I2	#I3
X_{CM} (m)	1.30	2.46	0.71	1.40	2.55	0.83	1.39	2.54	0.81
σ_{X} (m)	0.39	0.55	0.31	0.33	0.41	0.28	0.39	0.51	0.30
W_n (J)	0.98	1.51	0.79	0.16	0.26	0.11	0.60	0.99	0.39
W_t (J)	135.40	261.29	71.18	121.82	217.34	70.57	122.044	205.51	69.51
W_{bulk} (J)	14.34	17.72	13.29	6.97	9.86	5.14	7.51	10.21	5.89

Table 5.5. *Propagation of the mass center X_{CM}, extension/spreading σ_{X} and energies dissipated for MPM and DEM simulations when the friction coefficient at the base is set to 0.3 (i.e. an angle of friction of 16.7°).*

As for the cases #G3 and #I1 shown in Figure 5.2, results correlated properly for the other simulations (from #G1 to #I3). However, when using bricks instead of blocks in the DEM simulations, the spreading (σ_{X}) of the final deposit was reduced – see the vertical bars in Figure 5.3 for cases #H compared with cases #I. This kind of result is actually expected since the rotation of single elements is greatly dependent on its shape, being disfavored in this case with the use of bricks. But the continuum approach is not able to naturally catch it as the DEM does.

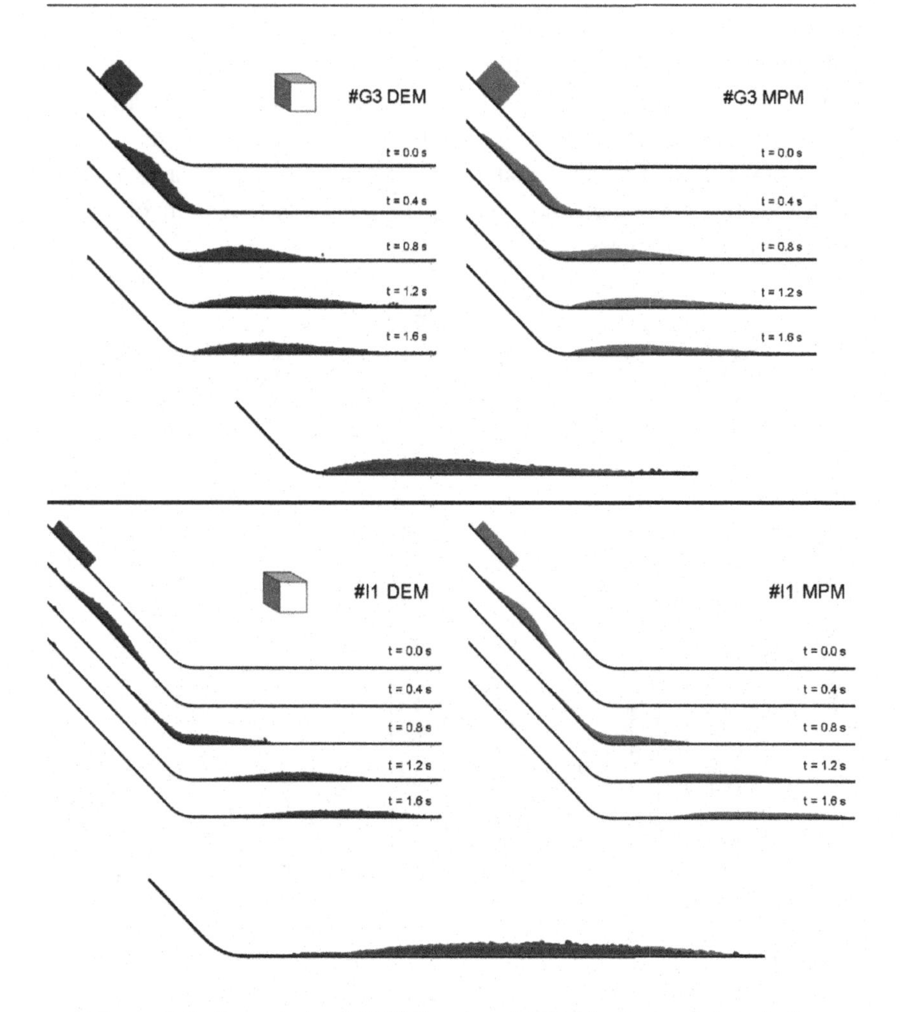

Figure 5.2. *Snapshots of DEM (blue) and MPM (red) simulations in the course of time for the configurations #G3 and #I1, and a friction coefficient at the base* $\mu_{\text{BS}} = 0.3$. *For a color version of this figure, see www.iste.co.uk/richefeu/gravity.zip*

In terms of energy dissipation, Figure 5.4 clearly shows that the driving means of dissipation is the friction at the base for both DEM and MPM simulations. This is actually expected since the internal friction angle (i.e. an input parameter of the constitutive model for the MPM, and a parameter mainly related to the friction between blocks in the DEM) is higher than the basal friction coefficient. Internal dissipation achieves basically the same values for both MPM and DEM. It is however slightly greater for higher container (cases #G). The worst (although good) matches appear for higher releases (cases #H2 and #I2) for which the constitutive model is longer involved to stretch the flowing mass.

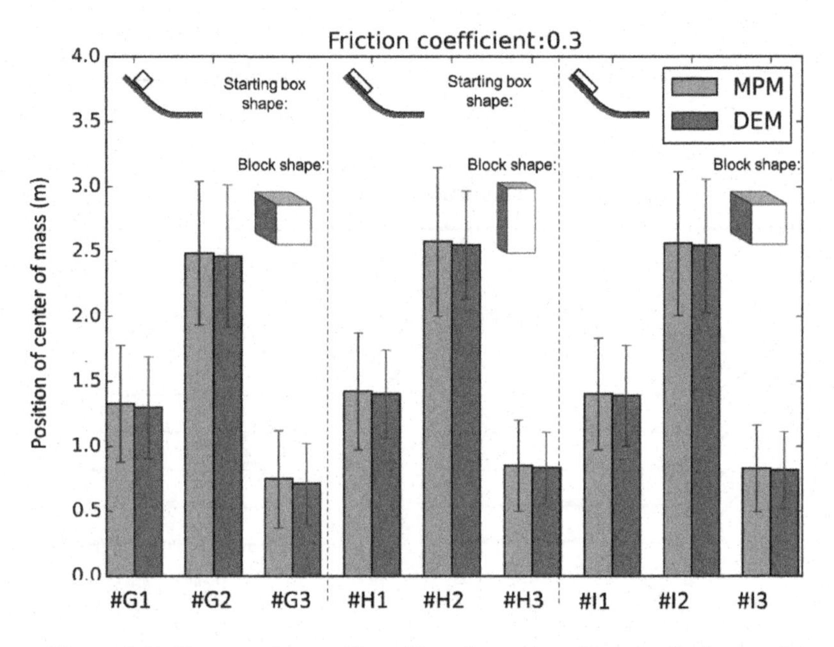

Figure 5.3. *Mass center positions X_{CM} (boxes) and longitudinal spread of the deposits (vertical bars) obtained by MPM and DEM simulations using a friction coefficient at the base $\mu_{\mathrm{BS}} = 0.3$*

The container shape becomes important, as well as the block shape, when analyzing the amount of energy dissipated within the granular mass. When using the square container, the energy dissipated in the bulk is larger. These differences might be attributed to the constitutive model itself or to the parameters used since they have not really be calibrated. In order to give a stronger role to the constitutive model, a second series of simulations were carried out with a higher friction at the base.

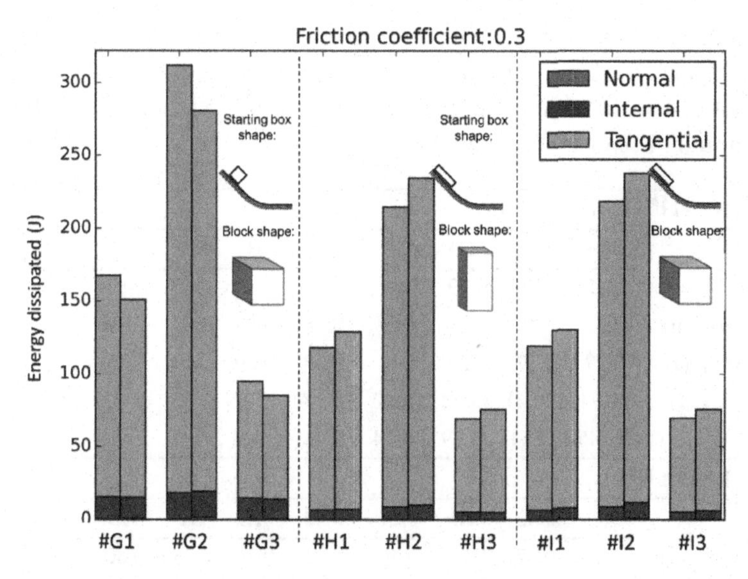

Figure 5.4. *Dissipated energy for MPM (left boxes) and DEM (right boxes) simulations. Three modes of dissipations are shown: collisions at the base (red), friction at the base (green) and internal plasticity (blue). The friction at the base was* $\mu_{\rm BS} = 0.3$*. For a color version of this figure, see www.iste.co.uk/richefeu/gravity.zip*

5.2.2. *A higher friction at the base*

The friction coefficient at the base was set to $\mu_{BS} = 0.5$ corresponding to an angle of 26.6°, which is still below but close to the internal angle of friction of 28.6°. Table 5.6 summarizes the results obtained. At first glance, it can be seen that similar results are found when comparing DEM and MPM in terms of propagation distance and spreading of the deposit, regardless of the aspect ratio of the starting box, release height or shape of blocks.

A closer look of the sequence of events is shown in Figure 5.5. Even if the match seems good, when regarding the position of mass center and the standard deviation around it, the shape of the final deposits is not exactly the same. In particular, the curvature near the slope transition differs between MPM and DEM simulations, and the scattering of the blocks in front of the deposit is completely missed by the MPM. Apart from that, the main mismatch seems to appear at the very end, when the final heap is forming. A more pronounced difference of volume can also be noticed with cubic container that induces more straining inside the continuum because of its flattening.

▶ **MPM,** $\mu_{BS} = 0.5$									
	#G1	#G2	#G3	#H1	#H2	#H3	#I1	#I2	#I3
X_{CM} (m)	0.45	0.76	0.29	0.47	0.82	0.30	0.46	0.81	0.30
σ_{X} (m)	0.32	0.40	0.27	0.29	0.38	0.24	0.29	0.36	0.23
W_n (J)	0.27	0.37	0.25	0.9	0.18	0.09	0.10	0.15	0.09
W_t (J)	90.71	229.41	33.42	78.40	185.74	37.16	80.29	174.32	36.96
W_{bulk} (J)	67.86	73.64	53.29	35.14	27.02	27.61	34.56	39.42	27.96
▶ **DEM,** $\mu_{BS} = 0.5$									
	#G1	#G2	#G3	#H1	#H2	#H3	#I1	#I2	#I3
X_{CM} (m)	0.42	0.77	0.25	0.45	0.81	0.26	0.45	0.80	0.27
σ_{X} (m)	0.28	0.41	0.22	0.24	0.32	0.20	0.25	0.37	0.19
W_n (J)	4.10	7.20	2.68	0.28	0.41	0.19	3.21	5.79	1.81
W_t (J)	99.82	208.90	46.31	114.32	214.78	64.12	94.23	186.45	50.65
W_{bulk} (J)	44.18	62.37	32.49	13.46	18.82	9.61	30.93	44.13	20.97

Table 5.6. *Propagation of the mass center X_{CM}, extension/spreading σ_{X} and energies dissipated for MPM and DEM simulations when the friction coefficient at the base is set to 0.5*

The graphical comparison of the deposit is shown in Figure 5.6. When compared with the results obtained with the low basal friction, the propagation distances and spreading of the deposit are lower for all configurations. The observations made with low basal friction are still valid with a basal friction angle close to the internal friction angle.

The partition of dissipation modes is different, as we can see in Figure 5.7. The part of the internal dissipation due to plastic straining has been increased. It is interesting to note that the best matches are obtained with a shallow container and cubic blocks for which the assumption of continuity is somehow more appropriate. In other configurations (in particular with bricks), the difference of internal dissipation is more pronounced because of the simplicity of the constitutive model. Despite this, the total amount of dissipated energy is rebalanced because of an increase in the friction dissipation at the base.

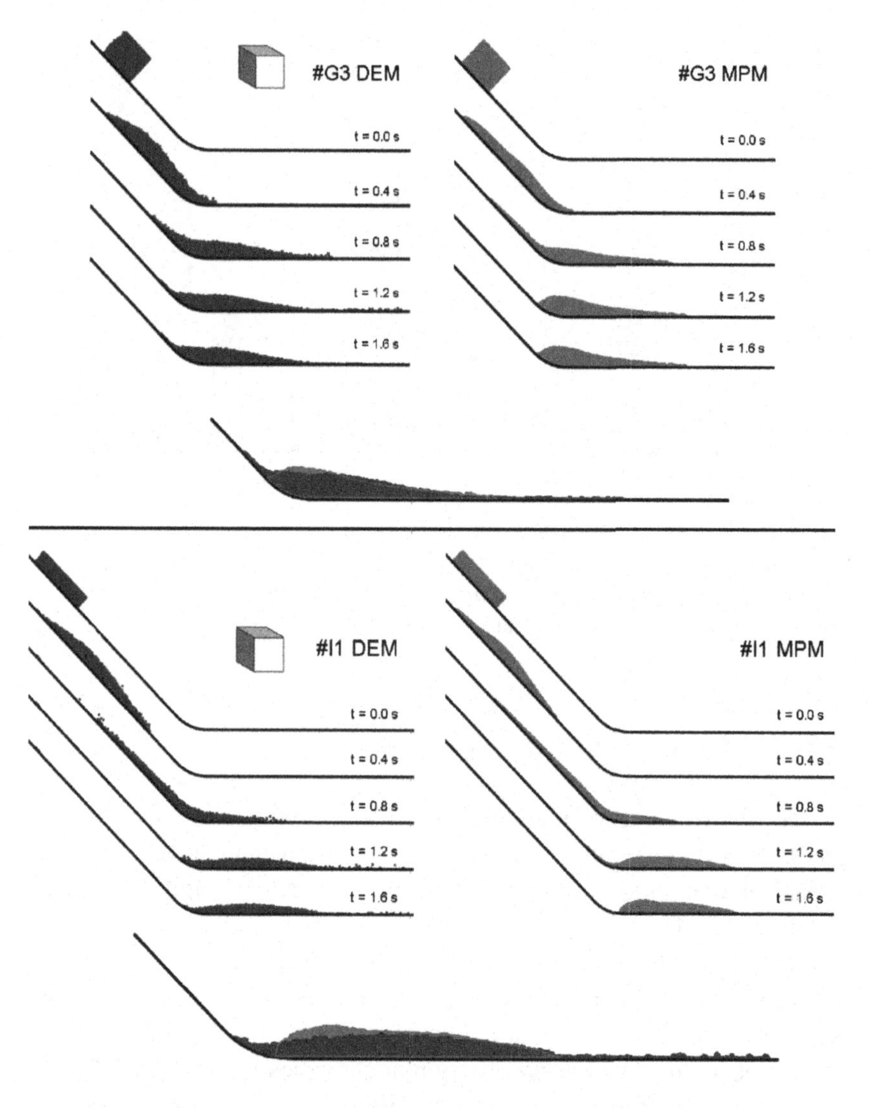

Figure 5.5. *Snapshots of DEM (blue) and MPM (red) simulations in the course of time for the configurations #G3 and #I1, and a friction coefficient at the base* $\mu_{\rm BS} = 0.5$. *For a color version of this figure, see www.iste.co.uk/richefeu/gravity.zip*

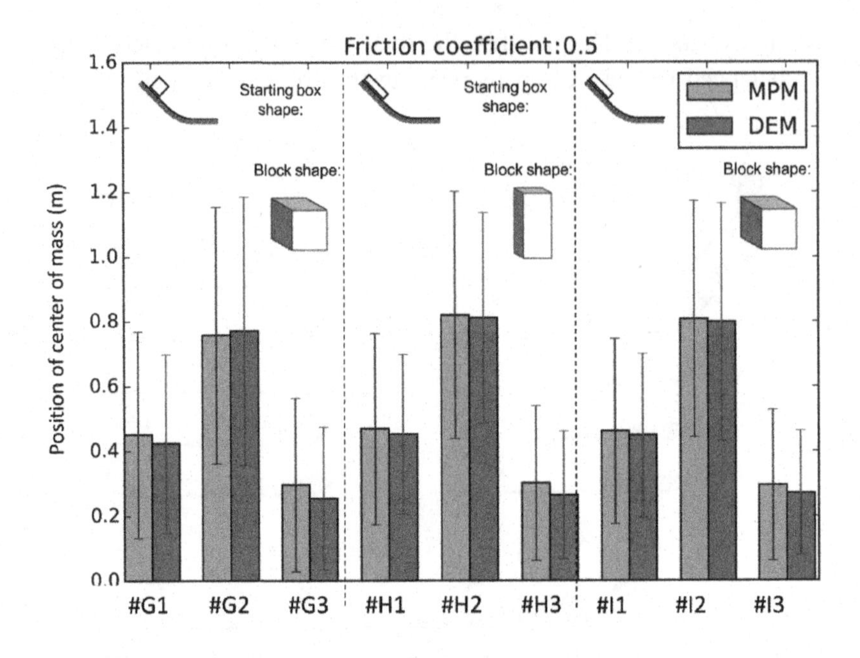

Figure 5.6. *Mass center positions X_{CM} (boxes) and longitudinal spread of the deposits (vertical bars) obtained by MPM and DEM simulations using a friction at the base $\mu_{\mathrm{BS}} = 0.5$*

5.2.3. *Sensitivity to the internal friction*

For the results shown above, the internal friction angle used in the elastoplastic constitutive law is always constant and lower than the basal friction. Since the Coulomb cohesion was negligible and the dilatancy angle was very small, the only plastic parameter was the internal friction angle. In order to analyze its effects, a number of additional simulations have been carried out. The common configuration makes use of a flat container, cubic blocks and a release height of 0.8 m. The internal friction angle varies from 10° to 80° for both low and high basal frictions. The influence of the internal friction angle on the propagation distance as well as the spread of the final deposit is shown in Figure 5.8.

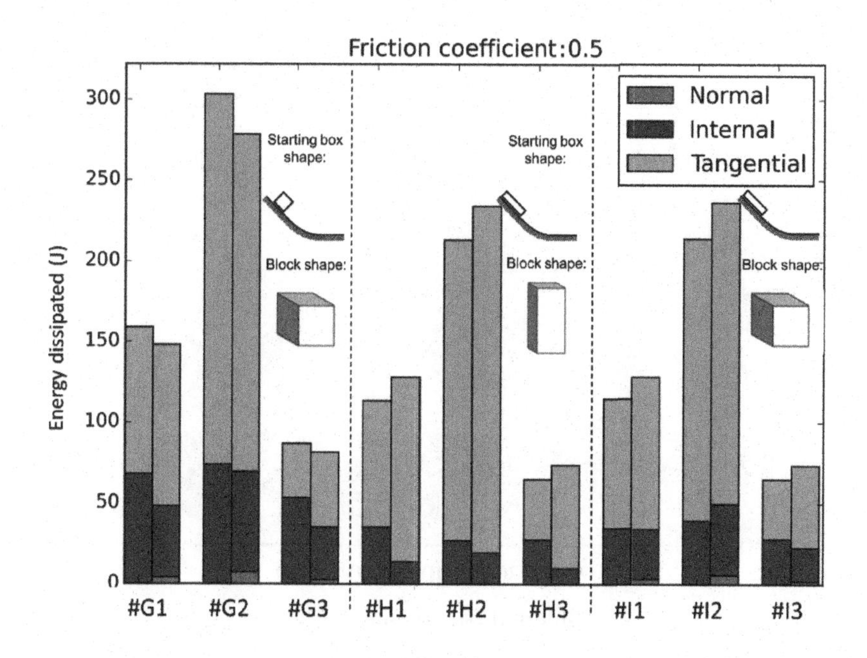

Figure 5.7. *Dissipated energy for MPM (left boxes) and DEM (right boxes) simulations. Three modes of dissipations are shown: collisions at the base (red), friction at the base (green) and internal plasticity (blue). The basal friction coefficient was* $\mu_{\mathrm{BS}} = 0.5$. *For a color version of this figure, see www.iste.co.uk/richefeu/gravity.zip*

Whatever the basal friction, a divergence of the propagation distance together with the mass spread is observed when the internal friction goes to zero. However, beyond an internal friction angle corresponding to friction angle $\tan^{-1}(\mu_{\mathrm{BS}})$ at the base, no marked change is noted. It is recalled here that these angles are, respectively, 16.7° and 26.6° for low and high basal frictions. This observation indicates that the release configuration – including the propagation terrain – allows for a simplistic model of plasticity to satisfactorily simulate the transient flow up to its stop. This would certainly not be true with some practical studies that can involve sophisticated digital surface models. In this case, an adequate model would obviously be unavoidable, but it is still impressive to see what the basic assumptions made here are able to predict.

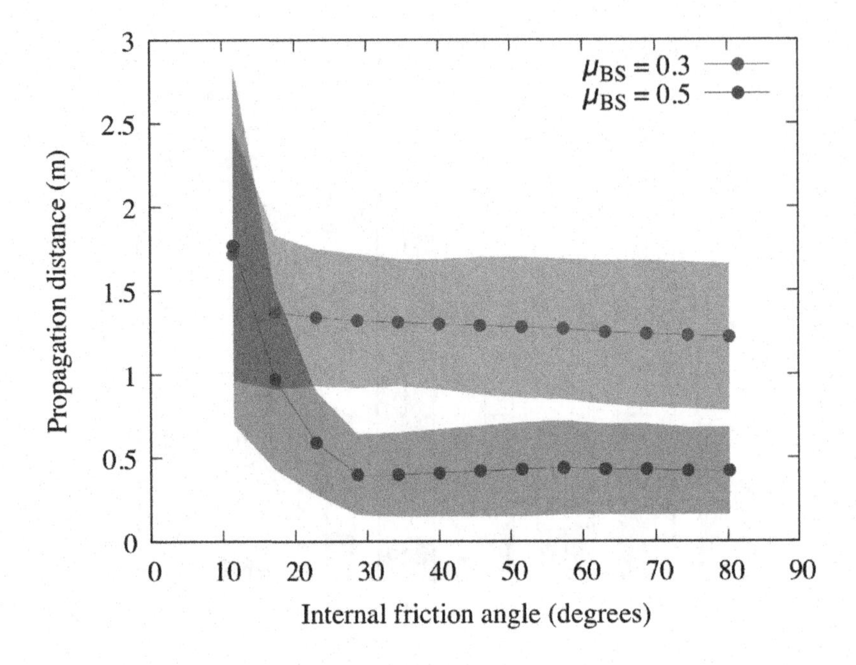

Figure 5.8. *Propagation distance X_{CM} as a function of the internal friction angle φ. The color surfaces are the spreading of the deposits $\pm\sigma_{\mathrm{CM}}$*

5.3. Concluding remarks

In this chapter, the use of the material point method has been evaluated in the difficult case of a flow in the transient regime that involves finite strains. In the comparisons presented, the DEM results were taken as the ground truth. Although an extreme simplification has been considered in the continuous model, MPM has proven to be of interest for modeling large size rock avalanches or landslides. In terms of energy dissipated, some discrepancies started to become evident as the friction coefficient was increased, but the results were still related.

In the field of soil (hydro-)mechanics or computational fluid dynamics, many constitutive models exist. MPM provides a framework that is not more restrictive than all approaches based on the continuum assumption. Accordingly, some events such as debris flows, mud flows or snow avalanches – deemed more complex to model – can be dealt with by implementing the best suited constitutive models. A more ambitious project is to make use of two-scale modeling by replacing the constitutive law by DEM simulations (representative to the material) for each material point.

Conclusion

Through the examples presented in this book, we showed the strong potential of the discrete element method to describe the propagation kinematics of interacting rock blocks and to assess the stop areas for rock avalanches and falling blocks. After validation by means of laboratory experiments, the implemented methodology has been extended to practical cases with certain success. Although the number of cases treated is not yet sufficient to assess the reliability of the model used, early results suggest interesting possibilities for its use for practical applications.

Among the model parameters, the block shape and the topology of the terrain have been shown to have the greatest impact on the kinematics of propagation, and consequently on the runout distances and morphology of the deposits. Even if angular shapes give rise to a variability in the bouncing of a single block, this effect is greatly reduced when considering a group of blocks that imply multiple and simultaneous interactions within the granular mass.

The main interaction parameters for introducing energy dissipation are the restitution coefficient in the direction normal to the contact, the tangential resistance coefficient that reflects the friction and abutment forces, and the rolling resistance coefficient that describes the energy loss by rotation experienced by a block when it collides with a soft soil. The latter coefficient is not really necessary for elongated particles, but it is absolutely required for more “spherical” blocks (including cubic shapes), especially when the slope is strongly inclined.

One advantage of the discrete element method lies in its ability to handle any kind of problem. The main limitations are the total number of blocks to consider or the fineness of the digital surface model, which induce very important – even prohibitive – computation duration. For applications involving large volumes, continuous models are preferred. They are significantly less expensive in computation time, and the assumption of continuity can reasonably be made. Another benefit of the discrete element method is that, given its formulation, it gives access to a large number of easily quantifiable or measurable data such as the modes of energy dissipation, the velocities at every point of the flow, the contact forces or collision energy (between blocks, with the terrain or with protective structures).

As with any numerical model, the attention given in the determination and the choice of the parameters is a key element to achieve realistic modeling. In the present case, the number of parameters remains very small and a procedure for determining the parameters has been proposed and validated in laboratory experiments. This procedure undoubtedly deserves to be extended to natural sites in order to improve the predictive nature of the numerical model. Later, it is hoped that a database of the cases handled can be established so that the sets of optimal parameters can be defined according to the nature and the topology of the terrain considered. Nevertheless, a certain expertise in using the model is required to optimize the computation duration and properly discretize the problem being treated.

In terms of perspective, we might think that, given the ever increasing evolving capacities of computation means, the computation duration can be reduced very quickly and that the method can be developed in the engineering world. Certain numerical developments are still possible to reduce the computation duration such as the optimization of the contact detection algorithms or the dissociation of the collective flowing phase from the trajectories of single ejected blocks.

Bibliography

[ALL 89] ALLEN M.P., TILDESLEY D.J., *Computer Simulation of Liquids*, Clarendon Press, New York, 1989.

[ALO 08] ALONSO-MARROQUIN F., "Spheropolygons: a new method to simulate conservative and dissipative interactions between 2D complex-shaped rigid bodies", *Europhysics Letters*, vol. 83, no. 1, 2008.

[BAN 09] BANTON J., VILLARD P., JONGMANS D. *et al.*, "Two-dimensional discrete element models of debris avalanches: parameterisation and the reproducibility of experimental results", *Journal of Geophysical Research Earth Surface*, vol. 114, 2009.

[BAR 02] BARDENHAGEN S.G., "Energy conservation error in the material point method", *Journal of Computational Physics*, vol. 180, pp. 383–403, 2002.

[BAR 04] BARDENHAGEN S.G., KOBER E.M., "The generalized interpolation material point method", *Experimental Arithmetic, High-Speed Computations and Mathematics*, vol. 5, no. 6, pp. 477–496, 2004.

[BER 03] VAN DEN BERGEN G., *Collision Detection in Interactive 3D Environments*, Morgan Kaufmann, San Francisco, 2003.

[BOT 14] BOTTELIN P., JONGMANS D., DAUDON D. *et al.*, "Seismic and mechanical studies of the artificially triggered rockfall at the Mount Néron (French Alps, December 2011)", *Natural Hazards and Earth System Sciences*, vol. 14, no. 2, pp. 3175–3193, 2014.

[BOU 09a] BOURRIER F., DORREN L., NICOT F. *et al.*, "Towards objective rockfall trajectory simulation using a stochastic impact model", *Geomorphology*, vol. 110, pp. 68–79, 2009.

[BOU 09b] BOURRIER F., ECKERT N., NICOT F. *et al.*, "Bayesian stochastic modeling of a spherical rock bouncing on a coarse soil", *Natural Hazards and Earth System Sciences*, vol. 9, pp. 831–846, 2009.

[CHR 07] CHRISTEN M., BARTELT P., GRUBER U., "RAMMS – a modelling system for snow avalanches, debris flows and rockfalls based on IDL", *PFG Photogrammetrie – Fernerkundung – Geoinformation*, vol. 4, pp. 289–292, 2007.

[CRO 04] CROSTA G.B., AGLIARDI F., "Parametric evaluation of 3D dispersion of rockfall trajectories", *Natural Hazards and Earth System Sciences*, vol. 4, pp. 583–598, 2004.

[CUE 15] CUERVO Y., Discret element modeling of rockfalls and application to real events, PhD Thesis, Université Grenoble Alpes, 2015.

[DAU 15] DAUDON D., VILLARD P., RICHEFEU V., *et al.*, "Influence of the morphology of slope and blocks on the energy dissipations in a rock avalanche", *Comptes Rendus Mécanique, Mechanics of Granular and Polycrystalline Solids*, vol. 343, no. 2, p. 166–177, 2015.

[DES 87] DESCOEUDRES F., ZIMMERMAN T., "Three-dimensional dynamic calculation of rockfalls", International Society for Rock Mechanics, Rotterdam, pp. 337–342, 1987.

[DOR 04] DORREN L., MAIER B., PUTTERS U.S., *et al.*, "Combining field and modelling techniques to assess rockfall dynamics on a protection forest hillstop in the European Alps", *Geomorphology*, vol. 57, pp. 151–167, 2004.

[FAL 85] FALCETTA J., "Un nouveau modèle de calcul de trajectoires de blocs rocheux", *Revue Française de Géomecanique*, vol. 30, pp. 11–17, 1985.

[HAR 64] HARLOW F.H., *The Particle-in-Cell Computing Method for Fluid Dynamics in Fundamental Methods in Hydrodynamics*, Academic Press, Cambridge, 1964.

[HAR 88] HART R., CUNDALL P., LEMOS J., "Formulation of a three-dimensional distinct element model – part 2: mechanical calculations for motion and interaction of a system composed of many polyhedral blocks", *International Journal of Rock Mechanics and Mining Sciences and Geomechanics Abstracts*, vol. 25, pp. 117–125, 1988.

[HUN 88] HUNGR O., EVANS S., "Engineering evaluation of fragmental rockfall hazards", *5th International Symposium on Landslides*, Lausanne, Switzerland/Balkema, Rotterdam, pp. 685–690, 1988.

[HUT 91] HUTTER K., KOCH T., "Motion of a granular avalanche in an exponentially curved chute – experiments and theoretical predictions", *Philosophical Transactions of the Royal Society London Series A-Mathematical Physical and Engineering Sciences*, vol. 334, pp. 93–138, 1991.

[HUT 95] HUTTER K., KOCH T., PLUSS C. *et al.*, "The dynamics of avalanches of granular-materials from initiation to runout. Part 2: experiments", *Acta Mechanica*, vol. 109, pp. 127–165, 1995.

[ITA 96] ITASCA CONSULTING GROUP I., Particles flow code in two dimensions, User manual, Minneapolis, 1996.

[LUD 03] LUDING S., TYKHONIUK R., TOMAS J., "Anisotropic material behaviour in dense, cohesive-frictional powders", *Chemical Engineering and Technology*, vol. 26, no. 12, pp. 1229–1232, 2003.

[MAN 08] MANZELLA I., LABIOUSE V., "Qualitative analysis of rock avalanches propagation by means of physical modeling of non-constrained gravel flows", *Rock Mechanics and Rock Engineering*, vol. 41, pp. 133–151, 2008.

[MAN 09] MANZELLA I., LABIOUSE V., "Flow experiments with gravel and blocks at small scale to investigate parameters and mechanisms involved in rock avalanches", *Engineering Geology*, vol. 109, 2009.

[MOL 12] MOLLON G., RICHEFEU V., VILLARD P. *et al.*, "Numerical simulation of rock avalanches: influence of a local dissipative contact model on the collective behavior of granular flows", *Journal of Geophysical Research – Earth Surface*, vol. 117, p. F02036, 2012.

[MOL 15] MOLLON G., RICHEFEU V., VILLARD P., *et al.*, "Discrete modelling of rock avalanches: sensitivity to size, aspect ratio and roundness of the blocks", *Granular Matter*, vol. 17, no. 5, pp. 645–666, 2015.

[MOR 03] MORESI L., DUFOUR F., MUHLHAUS H.-B., "A Lagrangian integration point finite element method for large deformation modeling of viscoelastic geomaterials", *Journal of Computational Physics*, vol. 184, pp. 476–497, 2003.

[NAI 03] NAIRM J.A., "Material point method calculations with explicit cracks", *CMES*, vol. 4, no. 6, pp. 649–663, 2003.

[OGE 98] OGER L., SAVAGE S.B., CORRIVEAU D. *et al.*, "Yield and deformation of an assembly of disks subjected to a deviatoric stress loading", *Mechanics of Materials*, vol. 27, pp. 189–210, 1998.

[PIT 76] PITEAU D.R., CLAYTON R., "Computer rockfall model", *Proceedings Meeting on Rockfall Dynamics and Protective Works Effectiveness*, Bergamo, Italy, pp. 123–125, 1976.

[PRE 07] PRESS W.H., TEUKOLSKY S.A., VETTERLING W.T. *et al.*, *Numerical Recipes: The Art of Scientific Computing*, 3rd ed., Cambridge University Press, 2007.

[RIC 12] RICHEFEU V., MOLLON G., DAUDON D. *et al.*, "Dissipative contacts and realistic block shapes for modelling rock avalanche", *Engineering Geology*, vol. 149, pp. 78–92, 2012.

[RIT 63] RITCHIES A., "Evaluation of rockfall and its control", *Highway Research Record*, vol. 17, pp. 13–28, 1963.

[SAV 91] SAVAGE S.B., HUTTER K., "The dynamics of avalanches of granular materials from initiation to runout. Part 1: analysis", *Acta Mechanica*, vol. 86, pp. 201–223, 1991.

[SOM 05] SOMFAI E., ROUX J.-N., SNOEIJER J.H. *et al.*, "Elastic wave propagation in confined granular systems", *Physical Review E*, vol. 72, p. 021301, 2005.

[SUL 94] SULSKY D., CHEN Z., SCHREYER H.L., "A particle method for history-dependent materials", *Computer Methods in Applied Mechanics and Engineering*, vol. 118, pp. 179–196, 1994.

[SUL 95] SULSKY D., ZHOU S.J., SCHREYER H.L., "Application of a particle-in-cell method to solid mechanics", *Computer Physics Communications*, vol. 87, pp. 236–252, 1995.

[SUL 96] SULSKY D., SCHREYER H.L., "Axisymmetric form of the material point method with applications to upsetting and Taylor impact problems", *Computer Methods in Applied Mechanics and Engineering*, vol. 139, pp. 409–429, 1996.

[TSU 92] TSUJI Y., TANAKA T., ISHIDA T., "Lagrangian numerical simulation of plug flow of cohesionless particles in a horizontal pipe", *Powder Technology*, vol. 71, no. 3, pp. 239–250, 1992.

Index

E, F, G

H, L, M

N, O, P

R, S

T, U, V

Printed in the United States
By Bookmasters